Personal Property
of
Bernarda Snyder

from: Kelly Huxel Ph.D

Anatomy and Human Movement Pocketbook

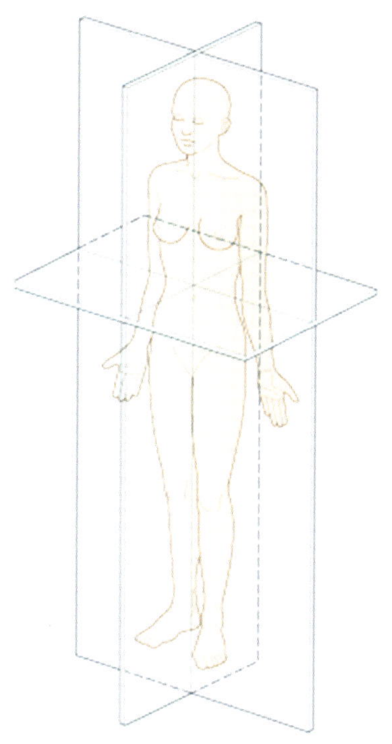

For Elsevier:

Publisher: Heidi Harrison
Associate Editor: Siobhan Campbell
Project Manager: Gail Wright
Senior Designer: George Ajayi
Illustration Buyer: Gillian Richards
Illustrator: Samantha Elmhurst

Anatomy and Human Movement Pocketbook

Nigel P. Palastanga MA BA FCSP DMS DipTP

Pro Vice Chancellor, Cardiff University (and Department of Physiotherapy Education), Cardiff, UK

Roger Soames BSc PhD

Centre for Anatomy and Human Identification, College of Life Sciences, University of Dundee, Dundee, UK

Dot Palastanga MA MCSP CertEd(HE)

Lecturer, Department of Occupational Therapy Education, Cardiff University, Cardiff, UK

CHURCHILL LIVINGSTONE

EDINBURGH LONDON NEW YORK OXFORD PHILADELPHIA ST LOUIS SYDNEY TORONTO 2008

CHURCHILL
LIVINGSTONE
ELSEVIER

First published 2008

ISBN 978-0-443-06912-3

British Library Cataloguing in Publication Data
A catalogue record for this book is available from the British Library

Library of Congress Cataloging in Publication Data
A catalog record for this book is available from the Library of Congress

Notice
Neither the Publisher nor the authors assume any responsibility for any loss or injury and/or damage to persons or property arising out of or related to any use of the material contained in this book. It is the responsibility of the treating practitioner, relying on independent expertise and knowledge of the patient, to determine the best treatment and method of application for the patient.

The Publisher

Printed in China

Contents

Preface

Nigel Palastanga and Roger Soames have been involved in the production of *Anatomy and Human Movement* (*AHM*) for almost 20 years. Now in its 5th edition, it has proved to be a very sucessful anatomy textbook. The detailed anatomy in *AHM* has necessarily made it a large and weighty volume. Students and clinicians have asked on many occasions for a smaller 'pocketbook' to accompany, but not replace, *AHM*. In response to this request the team has been joined by Dot Palastanga who had considerable involvement with all previous editions and is an experienced anatomy teacher for physiotherapists and occupational therapists.

The idea behind the pocketbook is to help those learning anatomy, particularly for vocational careers in the health sciences, by condensing the detailed text of *AHM* into more manageable segments. However, it is recognized that the detail provided in this book is insufficient for the level of anatomical knowledge required by a number of professions. It is envisaged that this book will be a useful aid to revision rather than the initial and only source of anatomical information. It is also anticipated that because of its size the pocketbook will be used by students and clinicians in the clinical setting, acting as a quick source of reference and revision. This is further supplemented by the study cards which are also available.

The illustrations in this book have been taken from *AHM* and in some cases amalgamated, especially showing functional groups of muscles related to the major joints. Finally, material is presented regionally centred on the major joints.

NPP/RS/DP

PART 1

Introduction

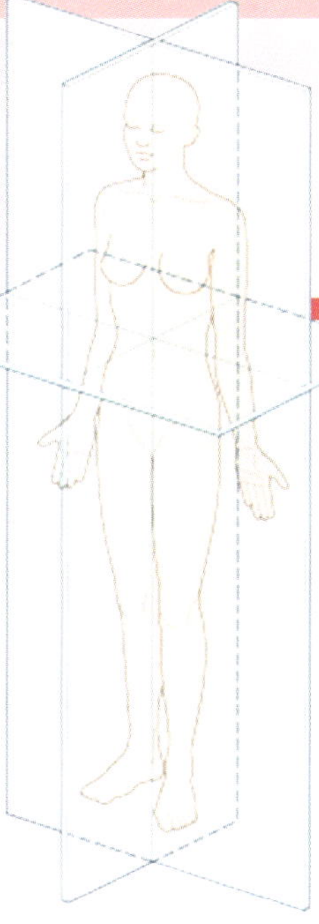

PLANES AND DIRECTIONS

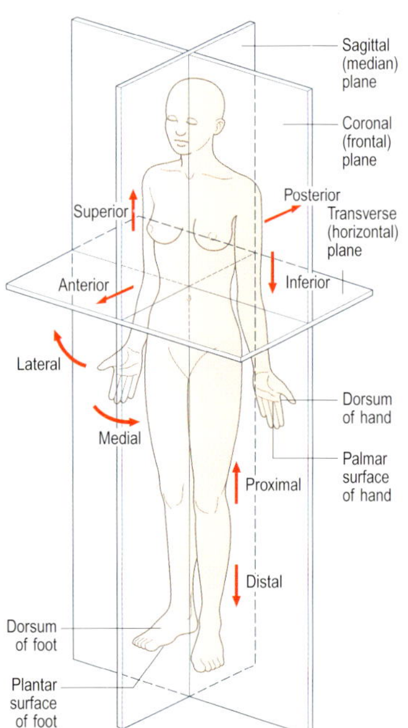

Anatomical position

Standing erect and facing forwards, lower limbs together, feet parallel and toes pointing forwards, upper limbs hanging loosely by the sides, palms facing forwards with the thumb lateral.

All planes, directions and movements are described with respect to the anatomical position irrespective of the actual position of the body.

Planes

Imaginary reference planes pass through the body in such a way that they are perpendicular to each other. A *sagittal (median) plane* passes through the body from front to back, dividing it into symmetrical right and left halves: parallel planes are *parasagittal planes*. A *coronal (frontal) plane* lies perpendicular to the sagittal plane, passing from top to bottom and dividing the body into anterior and posterior parts. A *transverse plane* lies perpendicular to the sagittal and coronal planes, dividing the body into upper and lower parts: the level of the transverse plane is given.

Directions

Directional terms describe the position of anatomical structures:

Anterior To the front or in front.
Posterior To the rear or behind.
Superior Above.
Inferior Below.
Lateral Away from median plane or midline.
Medial Towards the median plane or midline.
Distal Further away from trunk or root of limb.
Proximal Closer to trunk or root of limb.
Superficial Closer to surface of body or skin.
Deep Further away from body surface or skin.

MOVEMENTS

Movements usually occur in two or more planes simultaneously but it is convenient to describe them with respect to defined axes. Movement about a transverse axis in a parasagittal plane is flexion and extension; about an anteroposterior axis in a coronal plane is abduction and adduction; about a vertical axis in a transverse plane is medial and lateral rotation.

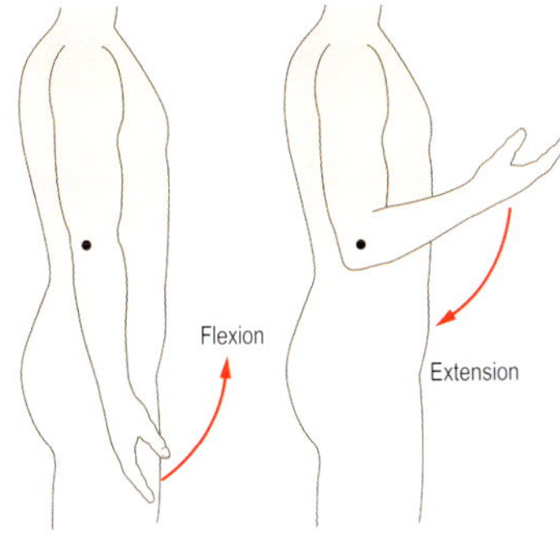

Flexion Bending of adjacent body segments so that their anterior, e.g. elbow, or posterior, e.g. knee, surfaces come together.

Extension The moving apart of two opposing surfaces, i.e. straightening a joint. Also refers to movement beyond the neutral position in a direction opposite to flexion.

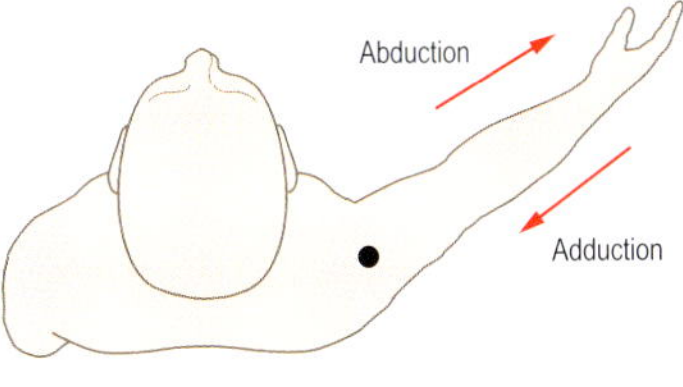

Abduction Movement of a body segment away from the midline of the body.

Adduction Movement of a body segment towards the midline of the body.

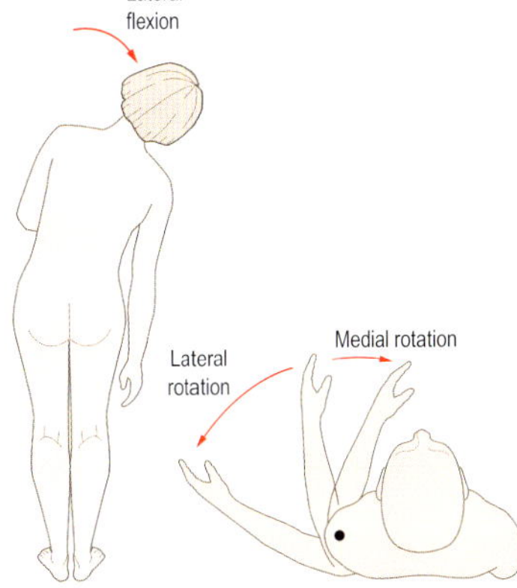

Lateral flexion (bending) Bending of the trunk to one side of the midline.

Medial rotation Rotation of a limb segment so the anterior surface faces towards the midline.

Lateral rotation Rotation of a limb segment so the anterior surface faces away from the midline.

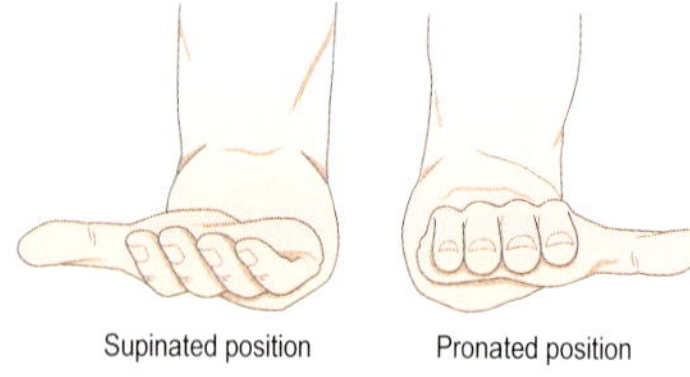

Movements of the forearm

Supination Movement of the forearm so the palm of the hand comes to face anteriorly.
Pronation Movement of the forearm so the palm of the hand comes to face posteriorly.

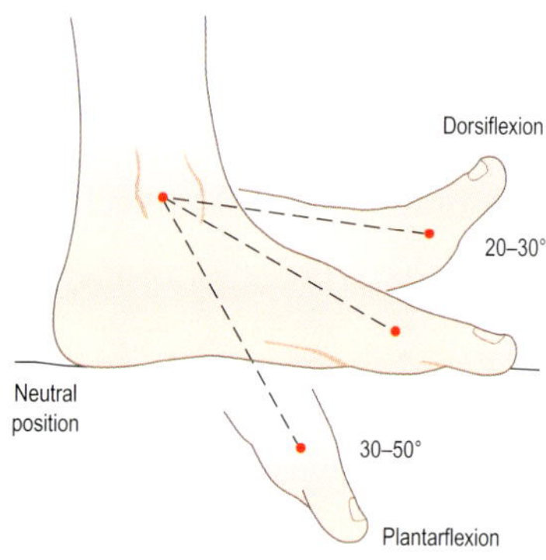

Movements of the foot

Plantarflexion Moving the dorsum (top) of the foot away from the anterior surface of the leg.
Dorsiflexion Moving the dorsum of the foot towards the anterior surface of the leg.

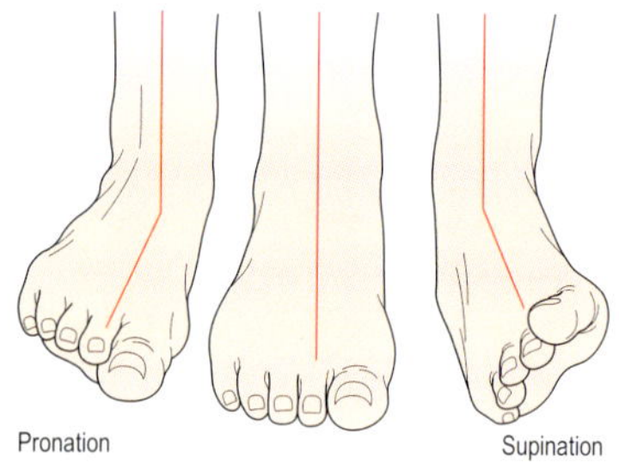

Supination Turning the forefoot so the sole faces medially: always accompanied by adduction of the foot.
Pronation Turning the forefoot so the sole faces laterally: always accompanied by abduction of the foot.

Inversion Movement of the whole foot to bring the sole to face medially: combined movement of supination and adduction.
Eversion Movement of the whole foot to bring the sole to face laterally: combined movement of pronation and abduction.

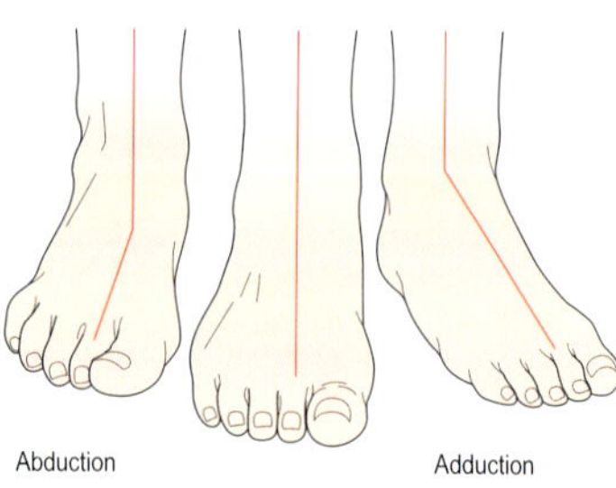

BONE

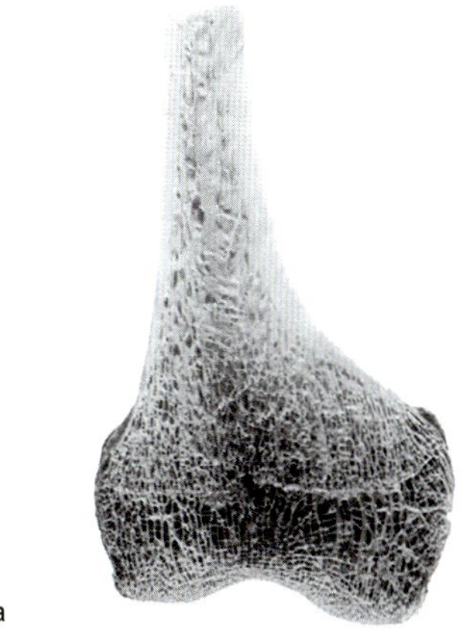

a

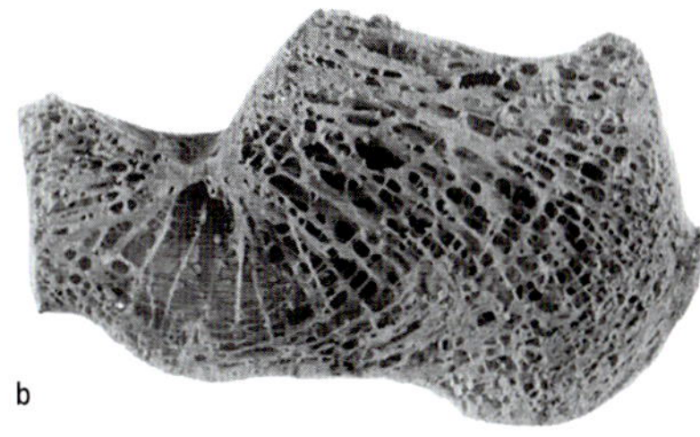

b

Compact and cancellous bone: (a) coronal section through shaft and distal end of the femur; (b) parasagittal section through the calcaneus

Bone is a composite of fibrous connective tissue, providing toughness and elasticity, and mineral salts, providing hardness and rigidity, surrounded by dense fibrous tissue (periosteum). It acts as a calcium store, which is continually exchanged with that in body fluids.

Bones are classified by their shape: *long bones*, which have a shaft (diaphysis) and two expanded ends (epiphyses); *short bones; flat bones; irregular bones.*

Cancellous (spongy) bone is found in the epiphyses of long bones, short and irregular bones where trabeculae (plates of bone) are arranged to resist compressive, tensile and shearing stresses. A thin layer of compact bone covers this cancellous bone and is found in the shafts of long bones.

Bone development

Intramembranous ossification is deposition of minerals in the mesoderm: it occurs in the skull, mandible and clavicle.

Endochondral ossification is the replacement of a cartilage model by bone: it occurs in all other bones.

Ossification centres are regions where ossification begins: primary centres appear before birth, secondary after birth.

Growth and remodelling

During growth bones change shape. In adults bone is continuously remodelled in response to circulating hormones and to long-term changes in applied forces. Growth and remodelling depend on the balanced activity of *osteoclasts* (bone-removing cells) and *osteoblasts* (bone-forming cells).

Growth in the length of long bones occurs between the diaphysis and epiphysis (*epiphyseal growth plate*): when this disappears growth ceases.

JOINTS

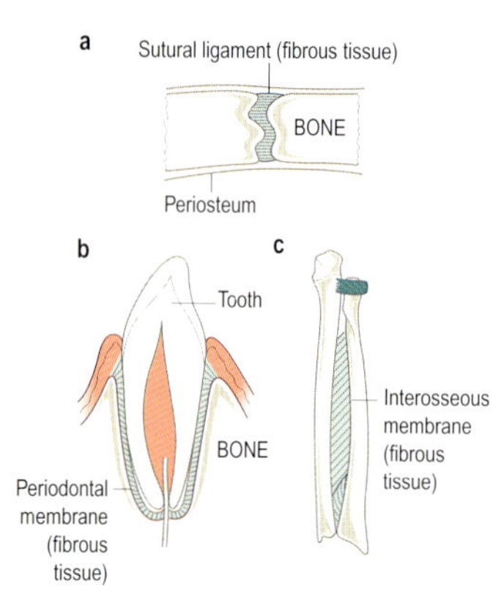

Fibrous joints: (a) suture; (b) gomphosis; (c) syndesmosis

In *fibrous joints* bones are connected by fibrous tissue and generally permit very little movement. There are three types: *sutures, gomphoses, syndesmoses.*

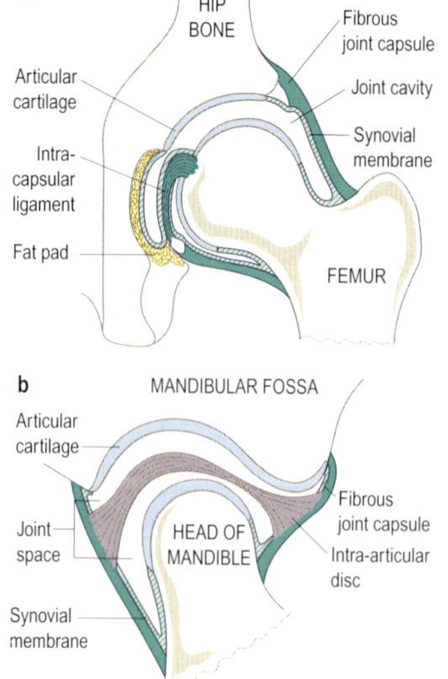

Synovial joint (a) without and (b) with an intra-articular disc

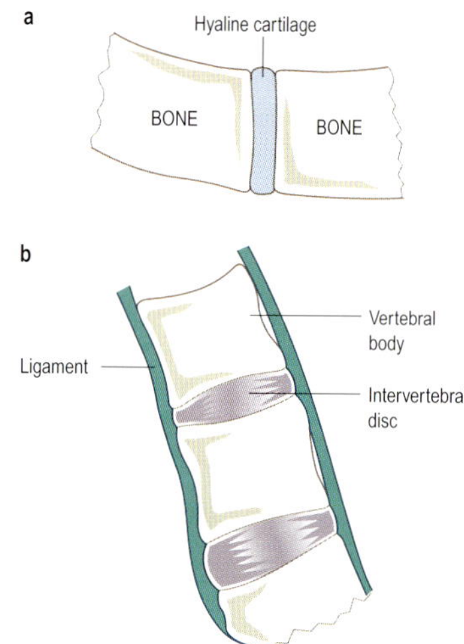

Cartilaginous joints: (a) primary; (b) secondary

In *cartilaginous joints* bones are united by a continuous pad of cartilage and allow limited movement. There are two types: *primary cartilaginous: secondary cartilaginous (symphysis).*

Synovial joints are freely mobile: surrounded by a fibrous capsule (often reinforced by ligaments), lined with synovial membrane which secretes synovial fluid into the joint space to aid lubrication. Bursae may be associated with synovial joints. The articular surfaces are covered in hyaline cartilage.

Synovial joints are subdivided according to the shape of the articular surfaces and movements possible at the joint: *plane, saddle, hinge, pivot, ball and socket, condyloid, ellipsoid.*

MUSCLE

Smooth muscle lines the walls of blood vessels and hollow organs: it contracts more slowly and less powerfully than skeletal muscle. *Cardiac muscle* forms the heart.

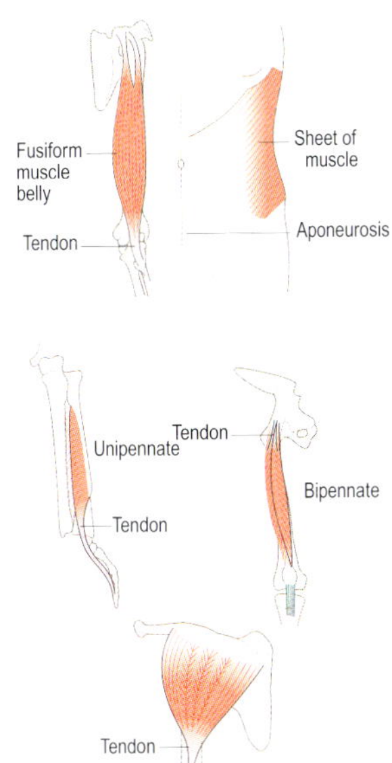

Arrangements of muscle fibres

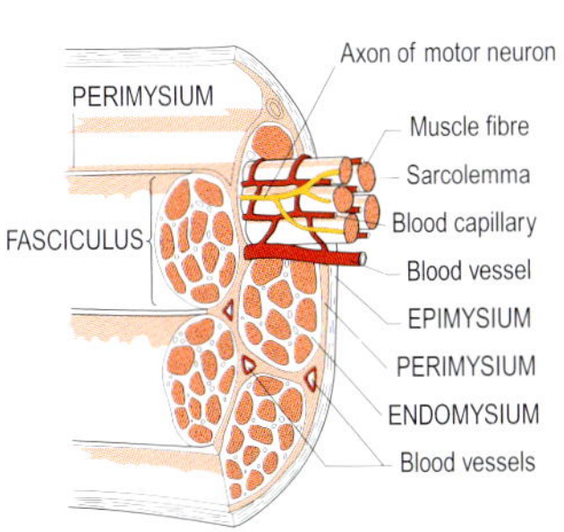

Organization of muscle

Skeletal muscle has individual fibres arranged either parallel or oblique to the line of pull of the whole muscle.

Parallel fibres form *fusiform* muscles or broad thin *sheets*, which can shorten considerably but have limited power.

Oblique fibres (*pennate*) cannot shorten to the same extent but are much more powerful. *Unipennate muscles* have fibres attached to one side of the tendon; *bipennate muscles* have fibres attached to both sides of a septum continuous with the tendon; *multipennate muscles* have intermediate septa, each with a bipennate arrangement of fibres.

When muscles contract they generate tension. If tension remains constant the contraction is *isotonic*: in *concentric* contraction the muscle shortens, in *eccentric* contraction the muscle lengthens. If muscle length remains unchanged (due to some externally applied force) the contraction is *isometric*.

Individual muscle fibres are surrounded by *endomysium* (connective tissue), bundles of parallel fibres (*fasciculi*) by *perimysium* and groups of fasciculi by *epimysium*.

Attachment of muscles to bone is by connective tissue in the form of a tendon, which may be round cords, flattened bands or thin sheets (aponeuroses).

Tendons subject to friction may develop sesamoid bones (e.g. patella), which also increase muscle leverage and change the direction of pull.

PERIPHERAL NERVES

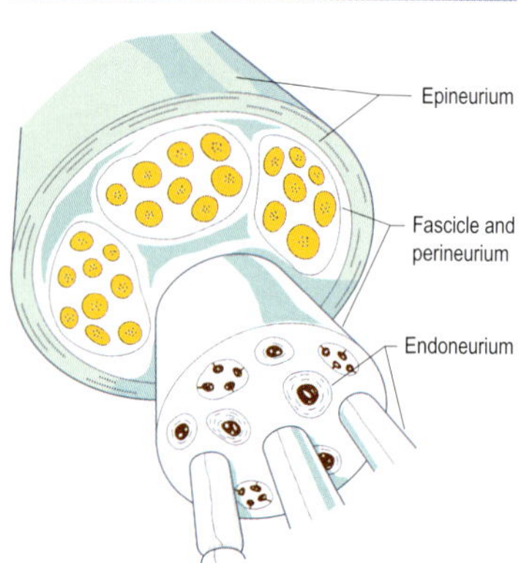

Peripheral nerve

Peripheral nerves consist of myelinated (surrounded by individual Schwann cell sheaths) and non-myelinated axons (embedded in invaginations of Schwann cell membrane).

Individual myelinated axons are surrounded by *endoneurium* (connective tissue), clusters of axons (*fascicles*) by *perineurium* and groups of fascicles by *epineurium*.

Afferent (sensory) fibres conduct impulses towards and *efferent (motor) fibres* away from the central nervous system. Sensory axons convey information about events in the periphery; motor axons cause events in the periphery (e.g. muscle contraction).

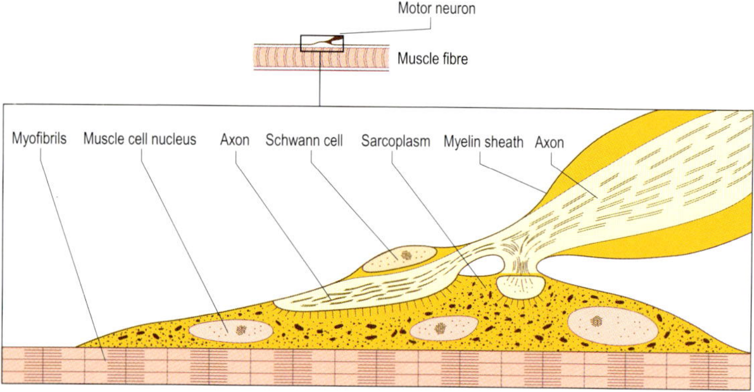

Neuromuscular junction

Neuromuscular junctions are complex regions between the terminal part of a motor axon and a muscle fibre (cell). As the nerve approaches a muscle cell it forms a flattened expansion against the surface muscle membrane: the junction is insulated from the surrounding environment by the Schwann cell sheath.

The terminal part of the nerve applied to the muscle cell is the *prejunctional membrane* and that part of the muscle cell membrane applied to the nerve is the *postjunctional membrane*: the two membranes are separated by a cleft. The terminal part of the nerve contains vesicles filled with neurotransmitter (acetylcholine) released when an action potential arrives at the nerve terminal. This crosses the cleft to react with receptors on the postjunctional membrane, generating an action potential in the muscle membrane leading to contraction of the muscle fibre.

LEVERS

The action and principles of levers are important in the consideration of forces applied to bones. A bone is a lever acting about a fulcrum (the joint about which the bone rotates): the distance between the joint and the muscle action (applied force) is the *force arm* and that between the joint and load (resistance) is the *load arm*.

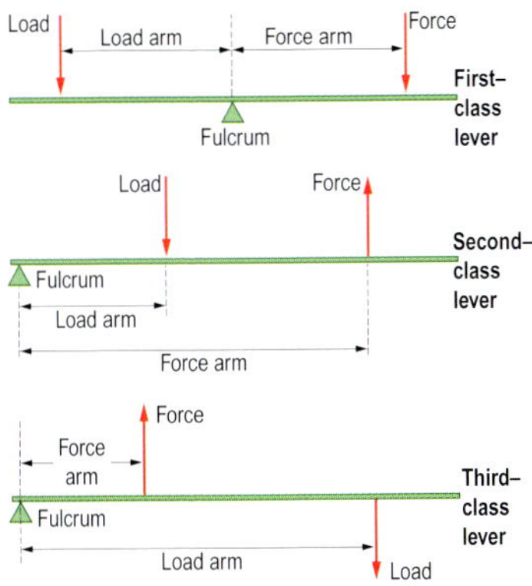

Levers

There are three classes of lever:

- *first-class levers* have the load and muscle action on opposite sides of the joint
- *second-class levers* have the load between the muscle action and the joint
- *third-class levers* have the muscle action between the load and the joint

All movement is dependent on the interaction between these three classes of lever.

The patella increases the leverage of the quadriceps muscle about the knee joint (a).

Biceps is acting as part of a third-class lever system (b). The major part of muscle action is directed towards stabilizing the joint rather than producing movement.

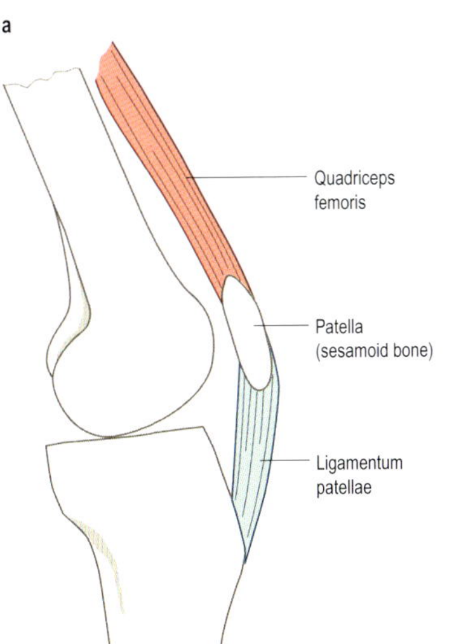

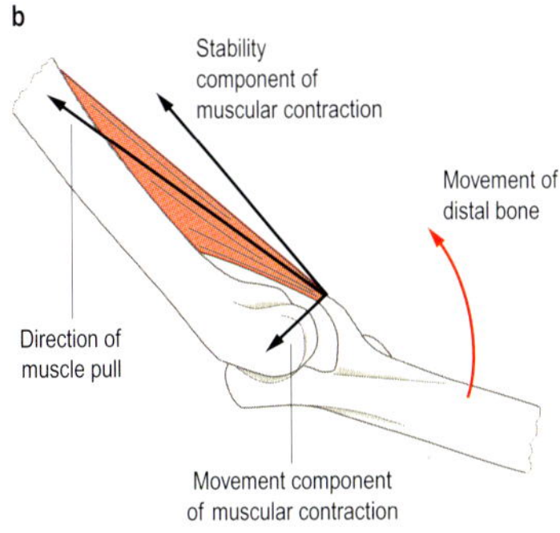

(a) Patella increasing muscle leverage; (b) biceps as third-class lever

SKIN

The superficial *epidermis* has several layers of cells which become keratinized and die as they move towards the surface. Its deep surface is firmly anchored to the underlying dermis and is avascular but contains sensory nerve endings. Epidermal melanocytes cause skin pigmentation.

The *dermis* contains blood vessels and lymphatic channels, nerves and their sensory endings, hair follicles, sweat and sebaceous glands, smooth muscle (arrector pili) and some fat. The deep surface is invaginated by subcutaneous tissue carrying blood vessels and nerves. Below the dermis is loosely arranged connective tissue containing blood and lymph vessels, roots of the hair follicles, secretory parts of sweat and sebaceous glands, cutaneous nerves and sensory endings, fat and some elastic fibres.

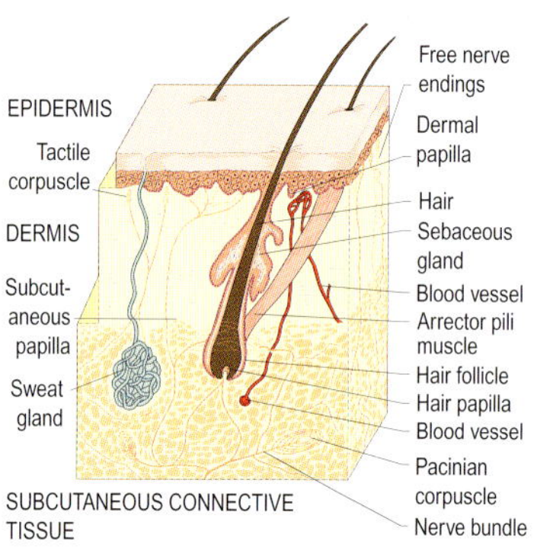

Skin

Nerve endings found in skin

- *Free nerve endings* detect stimuli generating pain and temperature: some also respond to touch and pressure
- *Merkel's discs* and *Meissner's corpuscles* mediate touch sensations
- *Krause's end bulbs* respond to mechanical stimulation of the skin
- *Ruffini corpuscles* respond to stretch when skin is deformed by pressure
- *Pacinian corpuscles* respond to pressure stimuli

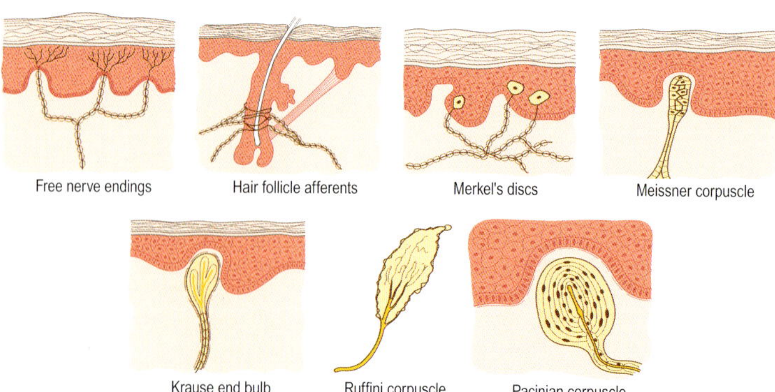

Nerve endings found in skin

LYMPHATIC SYSTEM

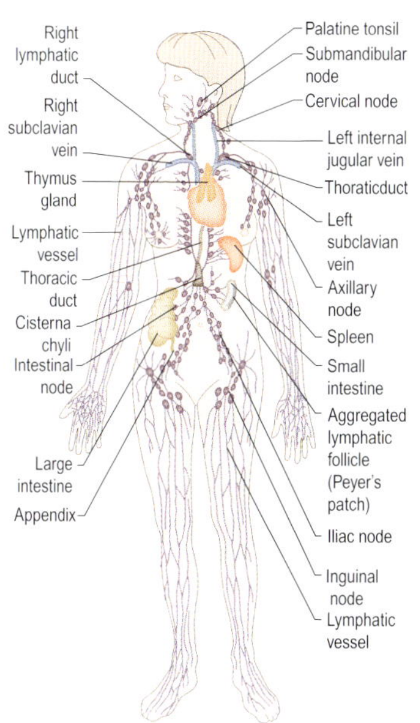

Anterior view of major lymph vessels and nodes

The lymphatic system is integrated closely with the vascular system.

Plasma from the circulatory system diffuses from capillaries and directly bathes the tissues. When this fluid is gathered up in the lymph vessels it is called *lymph* and returned to the heart. Composition of lymph is similar to blood but has lymphocytes instead of red blood cells. Larger vessels contain valves to help prevent backflow and along their course are groups of lymph glands called *nodes*.

APPLIED ANATOMY

Lymph glands act as filters, preventing passage of harmful bacteria/toxins from tissues to blood; in doing so they may become inflamed, swollen and painful.

The lymphatic system also helps regulate the amount of tissue fluid by absorbing excess and returning it to the main circulatory system. Excess fluid will produce *oedema* as in prolonged immobility or following radical mastectomy where axillary nodes are removed.

Flow in the lymphatic system can be improved by the following:

- Muscular movements around vessels which compress them while valves prevent backflow. This helps explain why muscular activity is so important to prevent oedema, particularly in the legs.
- Inspiratory movements of breathing help draw lymph into the thoracic duct and expiratory movements assist flow into major veins returning to the heart. Deep-breathing techniques are therefore important for good lymphatic flow.
- Massage strokes applied from distal to proximal in the limbs.
- Elevation of limbs above the level of axilla or groin will allow gravity to assist lymphatic drainage.

PART 2

The upper limb

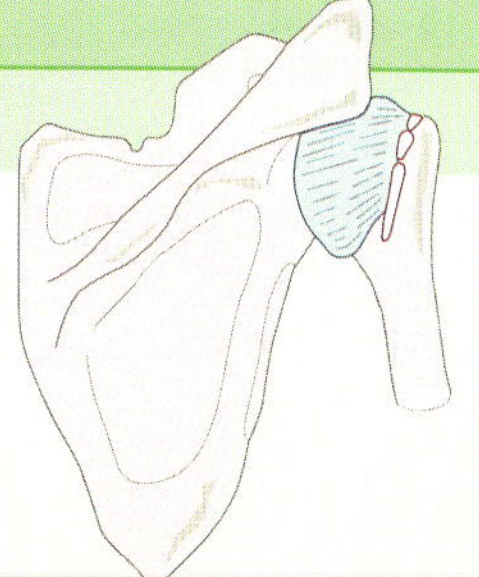

INTRODUCTION

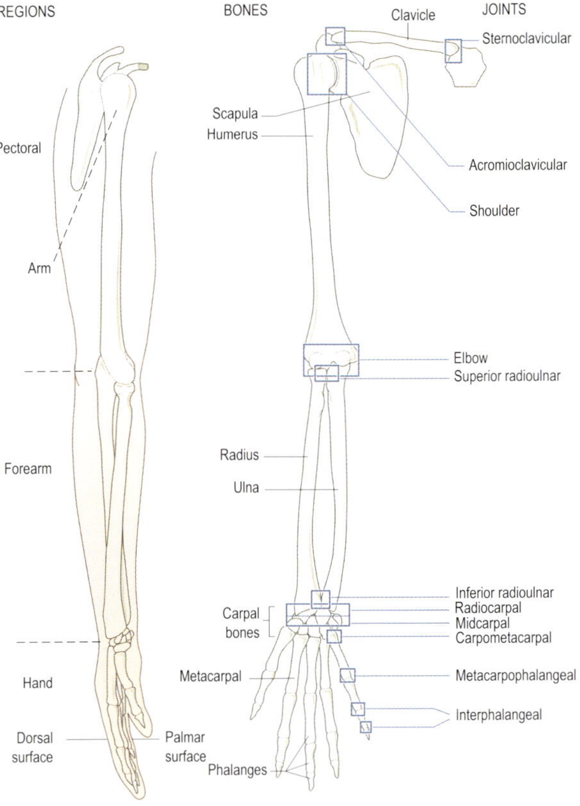

Evolutionary adaptation has resulted in the upper limb having almost no locomotor function but it has retained the ability to act as a locomotor prop as in using walking aids. The upper limb is attached to the trunk by the pectoral girdle where the scapula is held in a bed of muscle attaching it to the head, neck and thorax, with the clavicle acting as a strut holding the limb away from the trunk. Between the trunk and hand are a series of highly mobile joints which allow the hand to be positioned in space and held there whilst it performs a variety of complex tasks. It is the development of the hand, and particularly the opposable thumb which allows grasping and manipulative skills, that has made the hand the most efficient tool in the animal kingdom.

PECTORAL GIRDLE

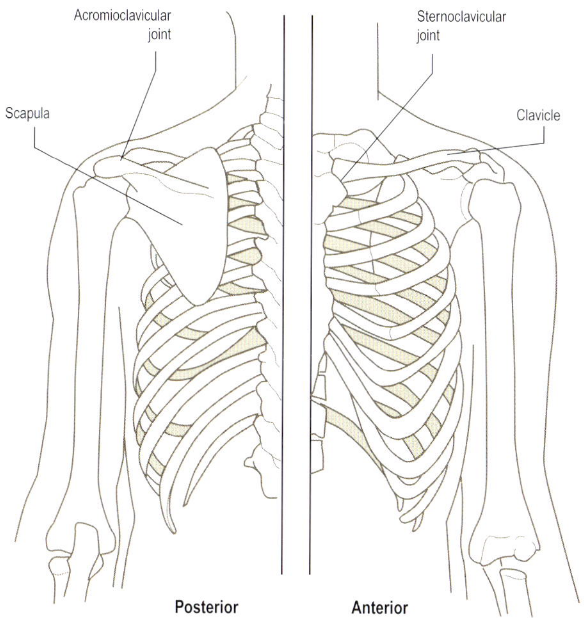

The pectoral (shoulder) girdle attaches the upper limb to the thoracic cage, providing a mechanism by which forces generated in the upper limb can be partially transferred to the axial skeleton without restricting movement of the limb as a whole.

The bones of the pectoral girdle are the *scapula* (shoulder blade), which lies over the 2nd to 7th ribs on the posterolateral aspect of the thorax, and the *clavicle* (collar bone). The scapula and clavicle articulate at the *acromioclavicular joint*: the medial end of the clavicle articulates with the manubrium of the sternum at the *sternoclavicular joint*. The scapula is attached to the thorax by muscles, while the clavicle acts as a strut holding the upper limb away from the trunk.

Combined movements of the sternoclavicular, acromioclavicular and shoulder (glenohumeral) joints allow the upper limb its considerable functional mobility. The clavicle moves with respect to the sternum, the scapula with respect to the clavicle and the humerus with respect to the scapula: in addition, the scapula moves with respect to the thorax. The great mobility of the shoulder–arm complex compromises stability but this is facilitated by powerful muscles attaching the pectoral girdle to the thorax, vertebral column, head and neck. This musculature also acts as a shock absorber when the upper limbs receive body weight.

BONES

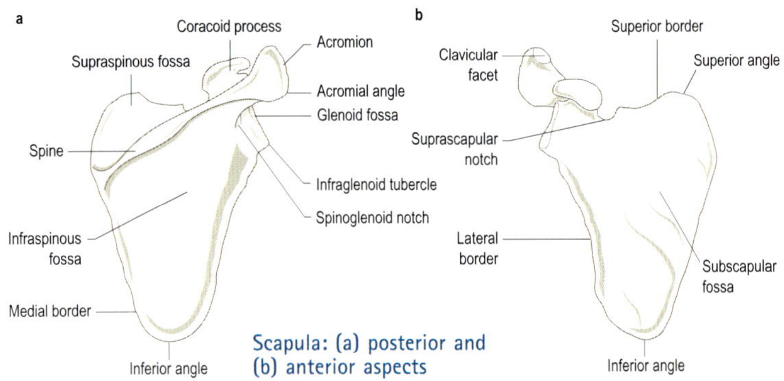

Scapula: (a) posterior and (b) anterior aspects

Scapula Flat triangular bone with two surfaces, three borders, three angles and two bony processes. The costal surface is smooth and ridged: the dorsal surface is divided by the spine into supra- and infraspinous fossae: the lateral angle is truncated as the glenoid fossa. The acromion process projects forwards from the spine: the coracoid process projects forwards below the clavicle.

Clavicle Subcutaneous bone between the sternum and the acromion; the medial two-thirds is triangular and convex forwards, the lateral third flattened and concave forwards. The superior surface is smooth; the inferior surface is marked by a large oval costoclavicular area medially and the conoid tubercle and trapezoid line laterally.

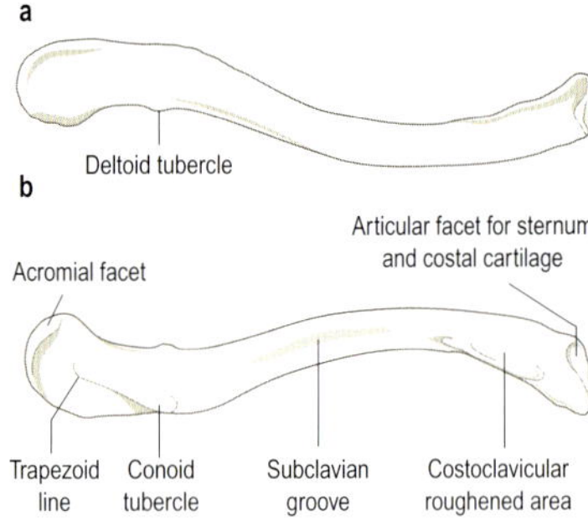

Clavicle: (a) superior and (b) inferior aspects

Palpation

Scapula The medial border can be followed along its whole length between inferior and superior angles. The spine starts as a small triangular area medially, increasing in size laterally: its lower border continues as the lateral border of the acromion. The coracoid process can be palpated below the middle and lateral third of the clavicle lying medial to the shoulder joint line.

Clavicle The enlarged medial end, sternoclavicular joint line, whole length of the shaft and acromioclavicular joint line can all be palpated.

ARTICULAR SURFACES

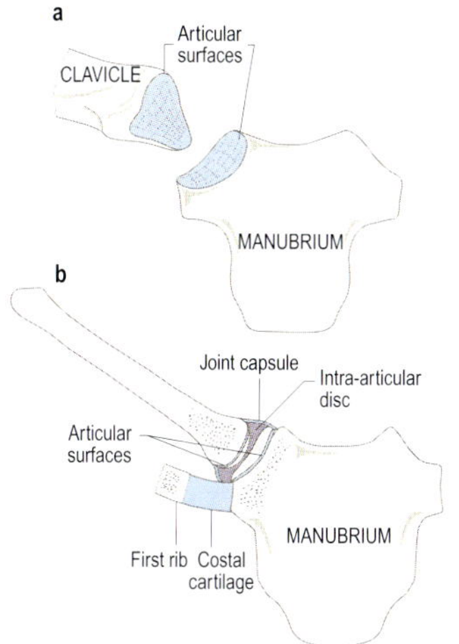

Sternoclavicular joint: (a) joint surfaces; (b) coronal section through joint

Sternoclavicular joint Medial end of clavicle and superolateral angle of manubrium of sternum and adjacent part of 1st costal cartilage: larger clavicular articular surface so that clavicle projects above manubrium. Articular surfaces covered with fibrocartilage.

Surfaces reciprocally concavoconvex but of different radii of curvature: congruence partly improved by fibrocartilaginous intra-articular disc.

Clavicular surface is convex vertically and flattened or slightly concave horizontally. Sternal surface at 45° to vertical is markedly concave from above downwards and convex from behind forwards.

Acromioclavicular joint Flat oval articular surfaces on lateral end of clavicle and anteromedial border of acromion: both surfaces covered with fibrocartilage.

Clavicular surface faces posterolaterally, acromial surface faces anteromedially.

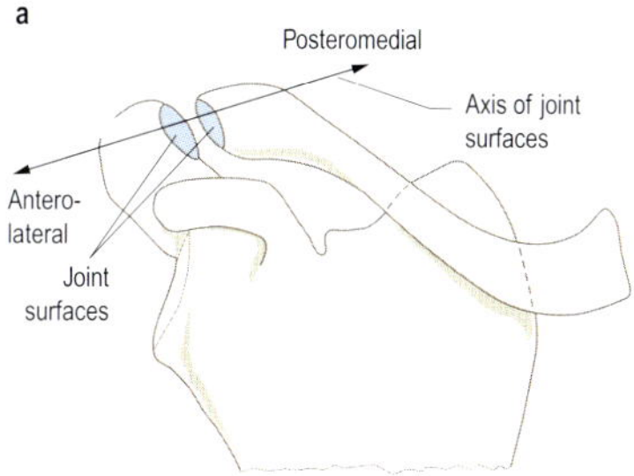

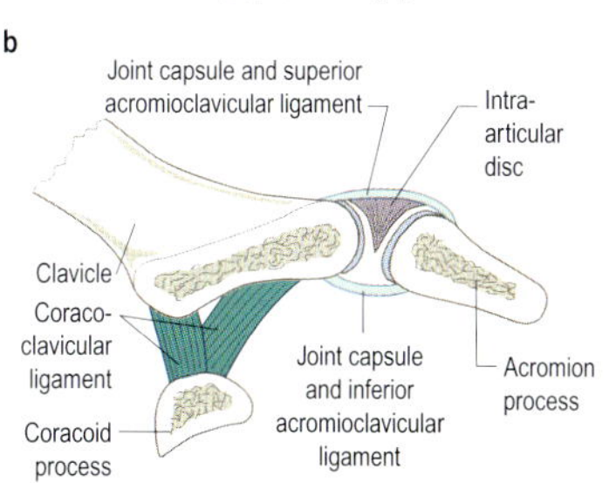

Acromioclavicular joint: (a) joint surfaces; (b) section through joint

CAPSULE AND LIGAMENTS

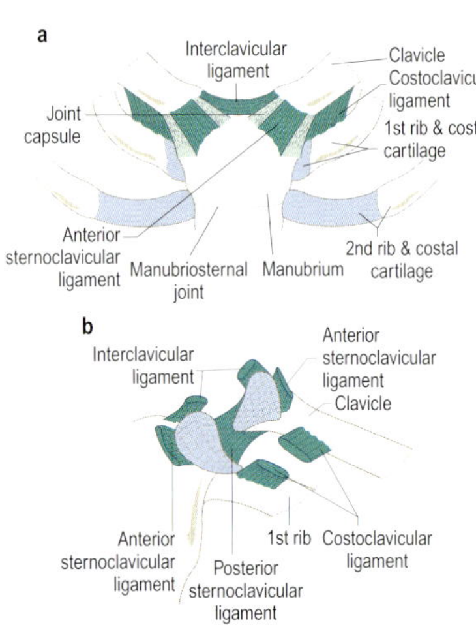

Sternoclavicular joint: (a) ligaments; (b) joint opened

Sternoclavicular joint Functionally a ball and socket joint. Strong *fibrous capsule* attaches to articular margins of clavicle and sternum, extending inferiorly on to upper surface of 1st costal cartilage. Reinforced by anterior, posterior and interclavicular ligaments. Complete intra-articular fibrocartilaginous disc attaches to capsule, dividing joint into two separate compartments: also attaches to posterosuperior border of medial end of clavicle and to sternal end of 1st costal cartilage. *Synovial membrane* lines each compartment.

Extracapsular *costoclavicular ligament* passes from upper surface of 1st costal cartilage to inferior surface of medial end of clavicle. It limits elevation of clavicle, as well as preventing excessive anterior or posterior movement of its medial end.

Acromioclavicular joint A synovial plane joint. Loose *fibrous capsule* attaches to articular margins reinforced by superior and inferior acromioclavicular ligaments. Partial fibrocartilaginous disc attaches to superior capsule in most joints. *Synovial membrane* lines all non-articular surfaces.

Extracapsular *coracoclavicular ligament* anchors lateral end of clavicle to coracoid process: conoid and trapezoid parts attach to conoid tubercle and trapezoid line respectively. Conoid part limits forward movement and trapezoid part backward movement of clavicle: both prevent medial displacement of acromion under the lateral end of the clavicle in laterally directed forces to the shoulder.

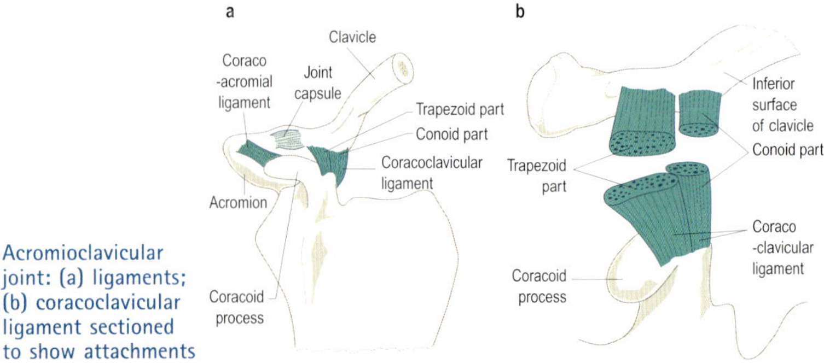

Acromioclavicular joint: (a) ligaments; (b) coracoclavicular ligament sectioned to show attachments

MOVEMENTS AT EACH JOINT

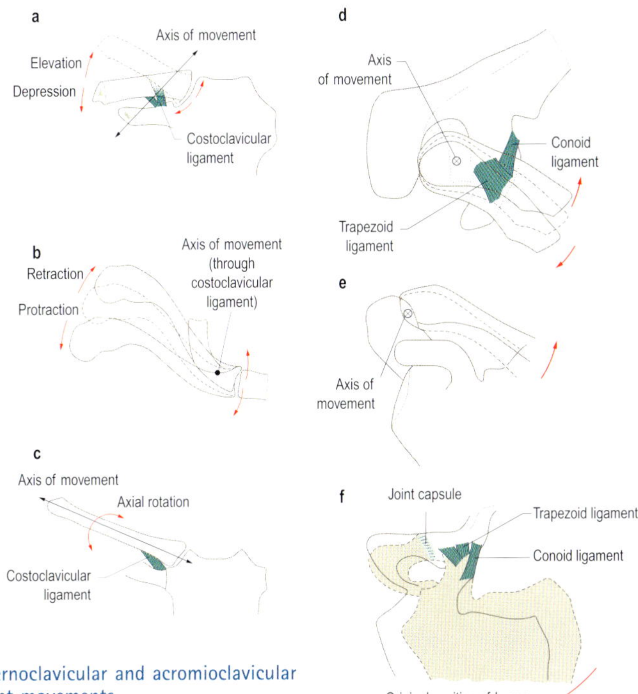

Sternoclavicular and acromioclavicular joint movements

Sternoclavicular joint

- *Elevation and depression* (a) about a horizontal axis through the costoclavicular ligament.
- *Protraction and retraction* (b) about a vertical axis through the costoclavicular ligament.
- *Axial rotation* (c) about an axis through the centre of sternoclavicular and acromioclavicular joints, being produced by rotation of scapula transmitted to clavicle by ligaments.

Acromioclavicular joint All movements are passive gliding of one articular surface against the other.

- *About a vertical axis* (d) is associated with protraction/retraction, with the angle between the clavicle and scapula increasing during retraction and decreasing during protraction.
- *About a sagittal axis* (e) is associated with elevation/depression.
- *Axial rotation* (f) is associated with medial and lateral scapular rotation occurring about an axis through the conoid ligament and acromioclavicular joint.

MOVEMENTS AS A WHOLE

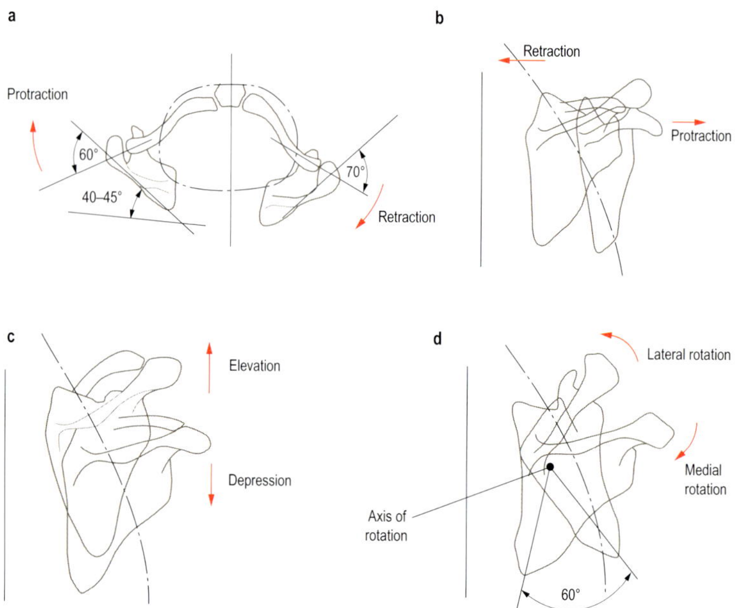

Scapular movements: (a) viewed from above; (b, c, d) posterior view right scapula

Movement of the pectoral girdle increases range of movement of the shoulder joint and thus upper limb.

Protraction/retraction (lateral/medial movement of the scapula around chest wall) brings the scapula to lie in a more sagittal or coronal plane respectively, with the angle between the extreme positions being 40–45°, the angle between the clavicle and scapula remaining between 60 and 70°. Linear translation of the scapula around the chest wall is about 15 cm.

Elevation/depression, usually accompanied by a small degree of scapular rotation, has a linear range of 10–12 cm.

Medial and lateral rotation occurs about an axis immediately below the spine close to the superomedial angle and has a range of 60°: this represents a linear displacement of the inferior angle of 10–12 cm.

During abduction or flexion of the arm the clavicle rotates about its long axis so that its superior surface becomes increasingly directed posteriorly. Restrictions to clavicular rotation will restrict movement of the upper limb as a whole.

MUSCLES

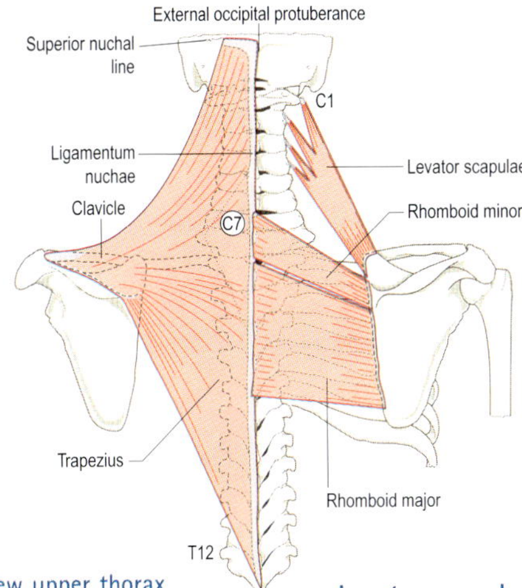

Posterior view upper thorax

Trapezius – retracts, elevates and laterally rotates pectoral girdle. Can also extend and laterally flex neck. Lower fibres can depress scapula.
Origin: medial third superior nuchal line and external occipital protuberance; ligamentum nuchae; spinous processes and supraspinous ligament C7–T12.
Insertion: posterior border lateral third of clavicle; medial border acromion; upper border spine of scapula.
Nerve supply: spinal accessory nerve (XI).

Levator scapulae – elevates, retracts and medially rotates pectoral girdle; laterally flexes and extends neck.
Origin: transverse processes C1–4.
Insertion: upper part medial border of scapula.
Nerve supply: dorsal scapular nerve C5.

Rhomboids major and minor – elevate, retract and medially rotate pectoral girdle.
Origin: both attach to spinous processes and covering ligaments: major – T2–5; minor – C7–T1.
Nerve supply: dorsal scapular nerve C5.

These muscles have an important role in stabilizing the scapula relative to the thorax during movements of the arm. Trapezius can also rotate the scapula laterally (with serratus anterior) to move the glenoid fossa and so increase range of elevation through abduction at the shoulder. From this position medial rotation to return the scapula to its normal position is usually eccentric work by the lateral rotators. However, medial rotation of scapula against resistance, as in pushing up out of a chair, is produced by rhomboids and levator scapulae (with pectoralis minor).

Trapezius (with levator scapulae) also has a postural function, maintaining the level of the shoulder and resisting downward gravitational pull on the arm, particularly when lifting or carrying weight.

MUSCLES

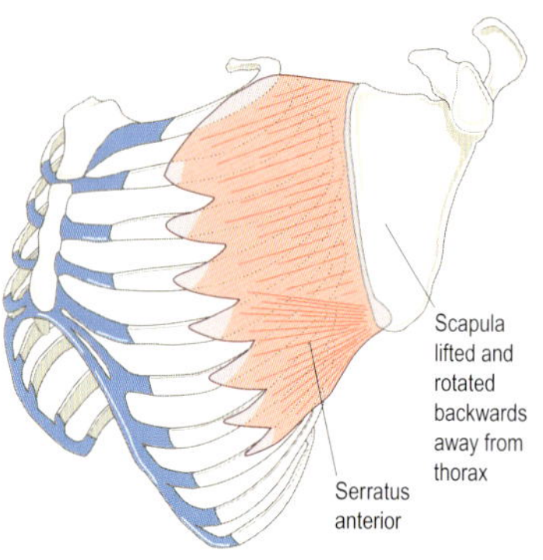

Oblique view of thorax

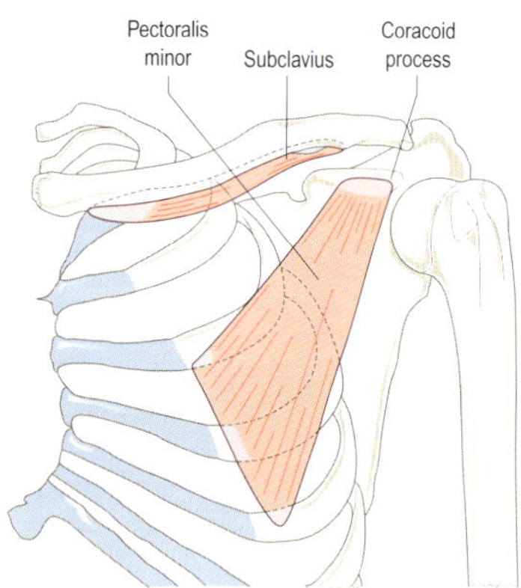

Anterior view left shoulder region

Serratus anterior – powerful protractor and lateral rotator of scapula.
Origin: outer surfaces upper 8 or 9 ribs and intercostal fascia.
Insertion: costal surface medial border of scapula.
Nerve supply: long thoracic nerve C5, 6, 7.

Serratus anterior is the major muscle involved in all pushing and punching movements where the scapula is driven forwards taking the upper limb forwards.

It is vital for stabilizing the scapula against the thorax during movements of the upper limb. Paralysis results in 'winging', severely affecting function and mobility.

It is an important lateral rotator of the scapula (with trapezius), increasing the range of abduction through elevation.

Pectoralis minor – lies deep to pectoralis major and is a protractor and medial rotator of the scapula.
Origin: coracoid process of scapula.
Insertion: outer surface ribs 3, 4, 5 and intercostal fascia.
Nerve supply: medial and lateral pectoral nerves C6, 7, 8.

Subclavius – stabilizes clavicle.
Origin and insertion: runs between 1st rib and undersurface of clavicle.
Nerve supply: nerve to subclavius C5, 6.

Pectoralis minor assists in medial rotation of the pectoral girdle and with the scapula fixed can elevate the ribs in forced inspiration.

ROTATION

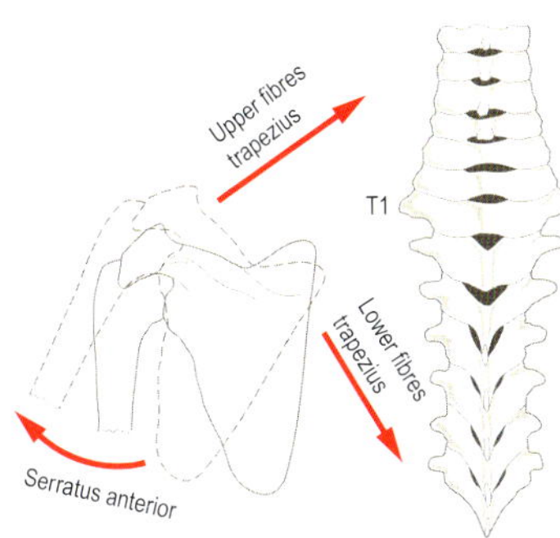

Lateral rotation of scapula (arrows indicate direction of pull of principal muscles involved)

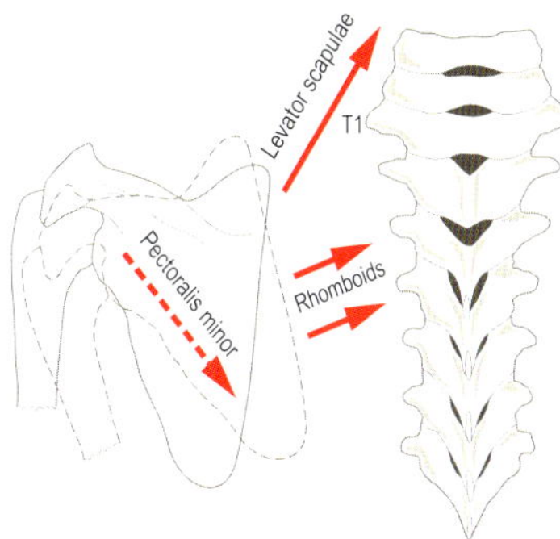

Medial rotation of scapula (arrows indicate direction of pull of principal muscles involved)

Lateral (forward) and medial (backward) rotation of scapula

Lateral rotation of the scapula is where the inferior angle moves laterally or outwards around the thorax accompanied by elevation of the lateral end of clavicle (these movements can be easily palpated). It is produced by serratus anterior pulling on the inferior angle, trapezius lifting the clavicle and acromion (upper fibres) and pulling down (lower fibres) on the medial end of the spine of scapula.

This movement will move the glenoid fossa of the shoulder, increasing the range of elevation through abduction by around 60°. The normal coordinated rhythm of this movement may become disrupted, giving a 'reversed humeroscapular rhythm'.

Medial rotation of the scapula is where the inferior angle of the scapula moves towards the spine accompanied by downward movement of the lateral end of clavicle (again movements which are easily palpated).

From the position of full lateral rotation, movement is often produced by gravity with the lateral rotators working eccentrically. Against resistance, e.g. when pushing up out of a chair, using crutches or performing a pull-up with arms fixed above the head, the medial rotators (levator scapulae, pectoralis minor and rhomboids) work concentrically.

SHOULDER

BONES

The *shoulder joint* is formed between the glenoid fossa of scapula and head of the humerus, and is a synovial ball and socket joint.

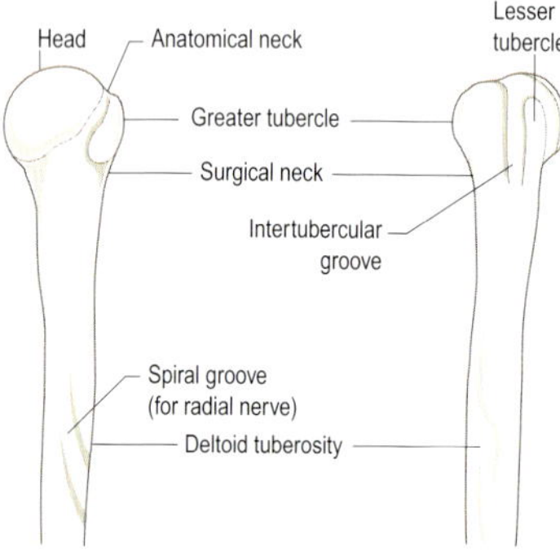

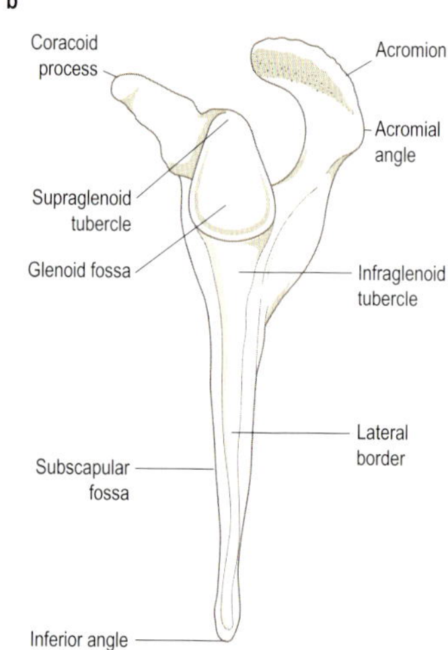

Shoulder joint: (a) proximal end of humerus; (b) lateral view of scapula

Mobility of the upper limb is partly due to changes in the shoulder joint which have occurred with freeing of the upper limb from weight-bearing and mobility of the pectoral girdle linking the upper limb to the trunk.

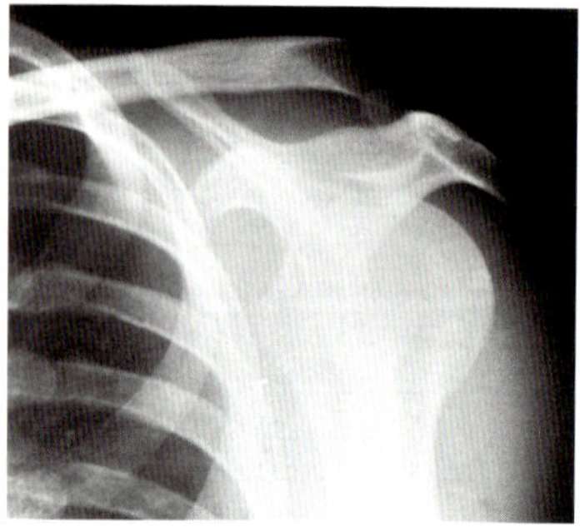

Radiograph of shoulder joint

Palpation

Scapula The upper surface of the acromion can be palpated above the shoulder joint, and the coracoid process anteriorly below the clavicle.

Humerus The greater tubercle is the most lateral bony point at the shoulder: the superior, anterior and posterior surfaces can be palpated. The rounded lesser tubercle can be felt through deltoid just lateral to the tip of the coracoid process. Lateral to the lesser tubercle the intertubercular groove can be felt: the upper part of the shaft can be palpated on its medial and lateral sides.

Shoulder joint line The surface projection of the shoulder joint line is represented by a vertical line slightly concave laterally, passing through a point 1 cm lateral to the tip of the coracoid process.

ARTICULAR SURFACES

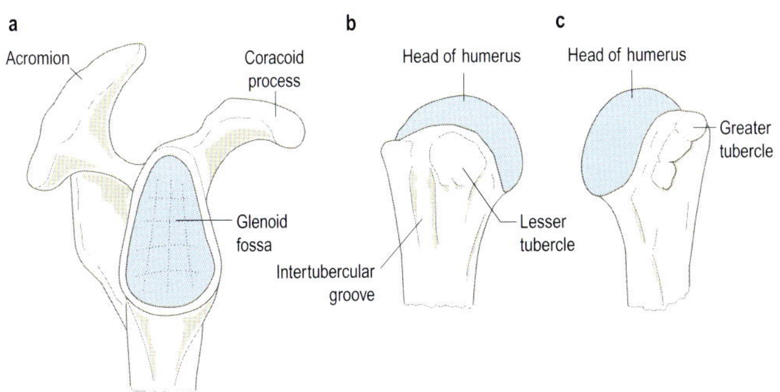

Shoulder joint articular surfaces: (a) glenoid fossa; (b, c) head of humerus

Medially the shallow pear-shaped glenoid fossa at the superolateral angle of the scapula faces laterally, anteriorly and slightly superiorly, being concave vertically and transversely. The articular surface is deepened by the presence of the fibrocartilaginous glenoid labrum but remains little more than one-third that of the humeral head.

Laterally the head of the humerus faces superiorly, medially and posteriorly, being two-fifths of a sphere. Irrespective of the position of the joint, only a third of the humeral head is in contact with the glenoid fossa at any one time, giving the joint its wide range of movement.

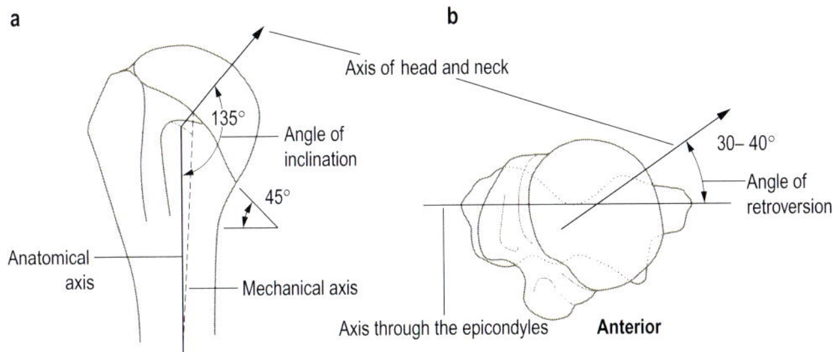

Axes associated with the proximal end of the humerus: (a) viewed anteriorly; (b) viewed from above

In the frontal plane the axis of the head and neck of the humerus forms an angle of 135–140° with the long axis of the shaft (angle of inclination): the centre of the head of the humerus lies approximately 1 cm medial to the long axis of the shaft. In addition, the axis of the head and neck is rotated backwards against the shaft 30–40° (angle of retroversion) so that the head faces posteromedially.

CAPSULE AND SYNOVIAL MEMBRANE

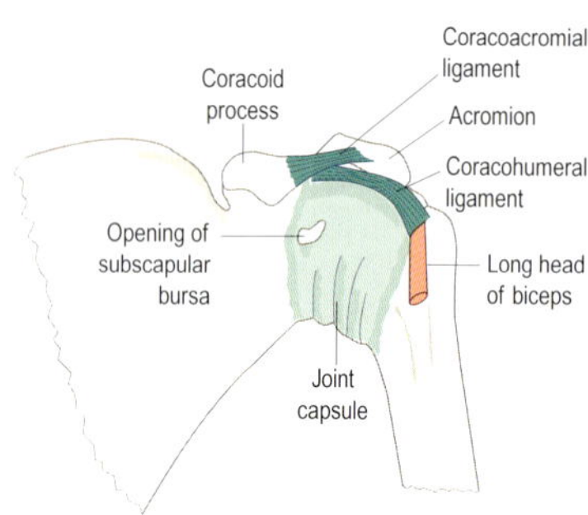

Anterior view of joint capsule

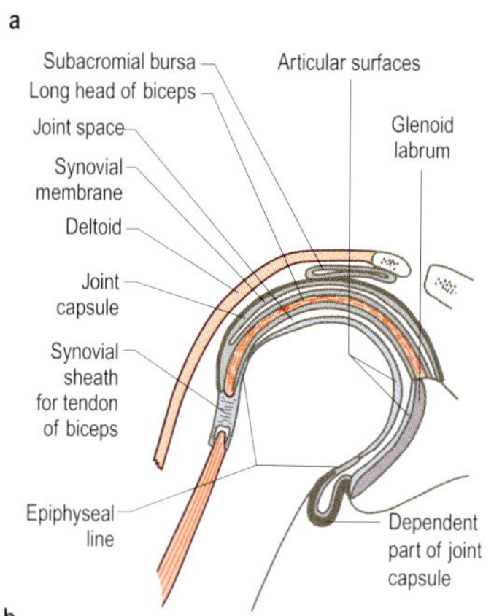

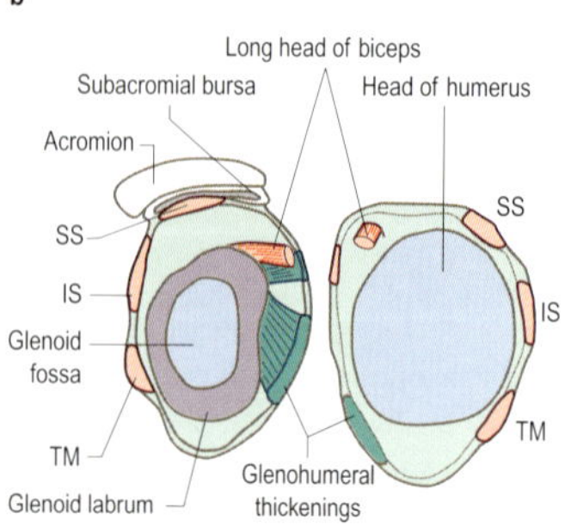

Articular surfaces: (a) coronal section through the shoulder joint; (b) the joint opened up showing the glenohumeral ligaments. IS – infraspinatus; SS – supraspinatus; TM – teres minor

Joint capsule A loose *fibrous capsule* completely surrounds the joint attaching to the glenoid labrum on the scapula and to the anatomical neck and articular margin of the humeral head, except inferiorly where it joins the upper part of the shaft 1 cm below the articular margin.

The anterior capsule is thickened by superior, middle and inferior glenohumeral ligaments. The rotator cuff tendons spread out over the capsule, blending with it near the humeral attachments. Although thick and strong in parts, the capsule conveys little stability to the joint. The tendon of long head of biceps passes through an opening in the capsule at the upper end of the intertubercular groove to enter the arm.

Synovial membrane Attaches to the articular margins of both bones covering all non-articular surfaces, being continuous with the subscapular and infraspinatus bursae. Other bursae (subacromial, subdeltoid), although clinically important, do not communicate with the shoulder joint space.

The intracapsular part of the long head of biceps is surrounded by a double-layered synovial membrane sheath continuous with that of the joint at its glenoid attachment, which passes deep to the transverse humeral ligament to enter the intertubercular groove extending 2 cm into the arm.

LIGAMENTS

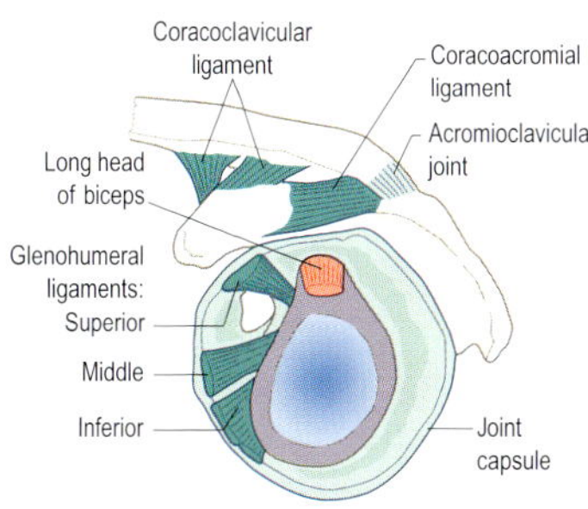

Lateral view of glenoid fossa, humerus removed

> Although the glenohumeral ligaments have no real stabilizing function they will be put under tension in lateral rotation of the shoulder, while in medial rotation they will become lax. In abduction the middle and inferior ligaments become taut and the superior lax.

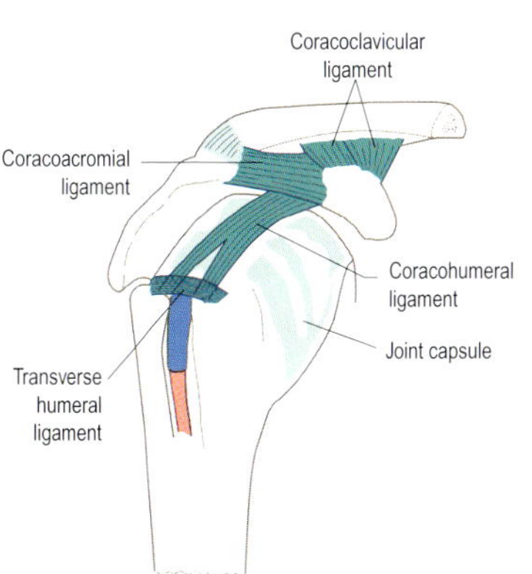

Anterior view

The anterior part of the joint capsule is reinforced by the superior, middle and inferior glenohumeral ligaments.

Superior glenohumeral ligament Passes from the upper glenoid margin and labrum to the upper surface of the lesser tubercle.
Middle glenohumeral ligament Arises below the superior ligament and passes to the front of the lesser tubercle.
Inferior glenohumeral ligament The best-developed ligament, passing from the glenoid margin to the anteroinferior part of the anatomical neck.

Transverse humeral ligament Bridges the gap between the greater and lesser tubercles at the upper end of the intertubercular groove.

Coracohumeral ligament A strong flat band from the lateral border of the coracoid process to the transverse humeral ligament and adjacent part of the anatomical neck.

Coracoacromial ligament A strong triangular ligament passing between the coracoid and acromion processes above the head of the humerus.

MOVEMENTS

Movement occurs about a number of axes passing through the centre of the humeral head: the principal axes must be carefully defined as the plane of the glenoid fossa does not lie in one of the cardinal planes (sagittal, coronal) of the body but at 45° to both. All movements involve a combination of gliding and rolling.

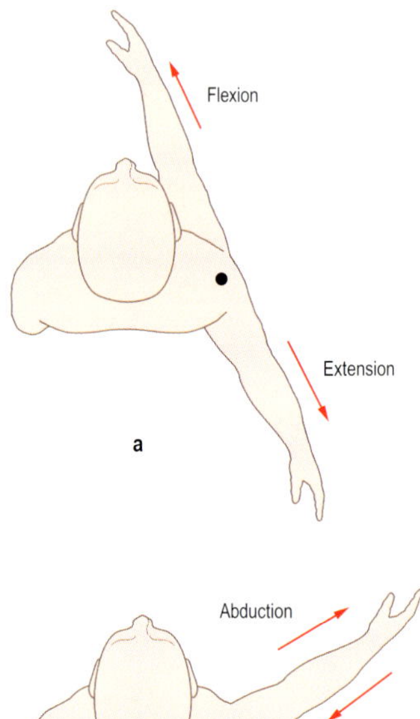

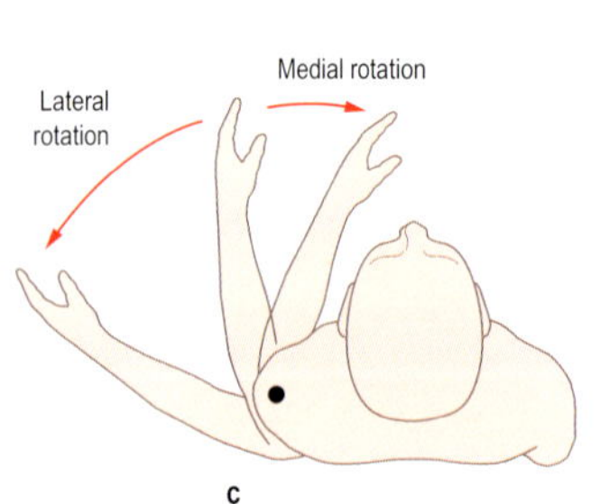

Movements at the shoulder joint with respect to the plane of the glenoid fossa: (a) flexion/extension; (b) abduction/adduction; (c) medial and lateral rotation

In movements with respect to the cardinal planes, flexion and extension occur about a transverse axis, abduction and adduction about an anteroposterior axis, and medial and lateral rotation about the long axis of the humerus.

In movements with respect to the plane of the glenoid fossa (as shown here) flexion and extension occur about an axis perpendicular to the fossa, abduction and adduction about an axis parallel to the fossa, and medial and lateral rotation about the long axis of the humerus.

- *Flexion and extension* (a) have a range of movement of 110° and 70° respectively.
- The range of *abduction* (b) is 120°, with only the initial 25–30° occurring without concomitant rotation of the scapula.
- *Lateral* and *medial rotation* (c) have a range in excess of 80° and 90° respectively.

Accessory movements: Proximal and distal gliding can be produced by forces applied along the shaft of the humerus: anterior and posterior gliding of the humeral head against the glenoid fossa can be produced by applying pressure to the region of the surgical neck.

MUSCLES

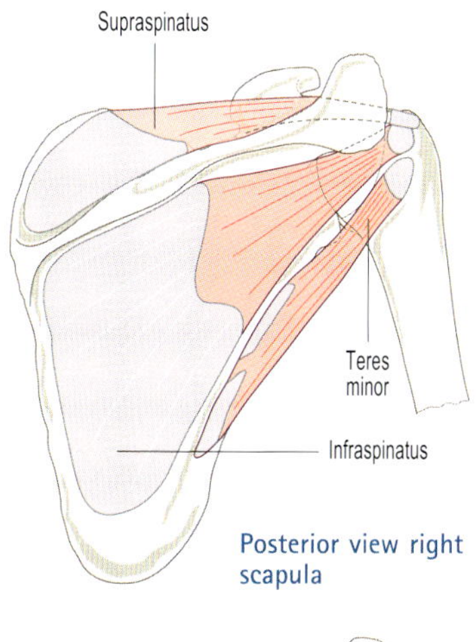

Posterior view right scapula

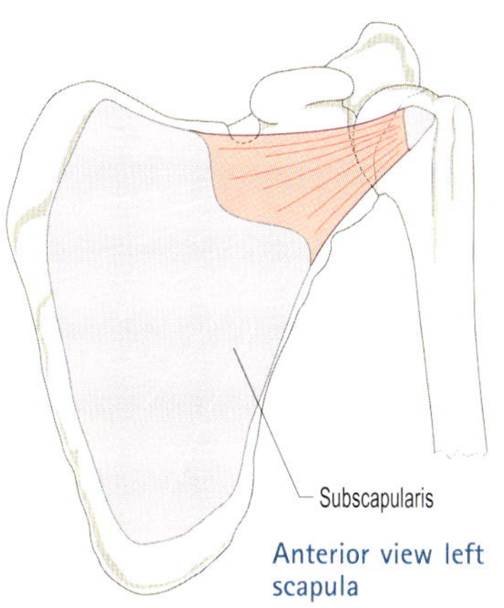

Anterior view left scapula

Supraspinatus – abducts arm at shoulder joint.
Origin: medial two-thirds supraspinous fossa and covering fascia.
Insertion: upper facet greater tubercle of humerus.
Nerve supply: suprascapular nerve C5, 6.

Infraspinatus – laterally rotates arm at shoulder joint.
Origin: medial two-thirds infraspinous fossa of scapula and covering fascia.
Insertion: middle facet greater tubercle of humerus.
Nerve supply: suprascapular nerve C5, 6.

Teres minor – laterally rotates arm at shoulder joint.
Origin: upper two-thirds lateral border of scapula.
Insertion: inferior facet greater tubercle of humerus.
Nerve supply: axillary nerve C5, 6.

Subscapularis – medially rotates arm at shoulder joint.
Origin: medial two-thirds subscapular fossa and covering fascia.
Insertion: lesser tubercle of humerus.
Nerve supply: upper and lower subscapular nerve C5, 6, 7.

These four muscles form the musculotendinous or rotator cuff which is essential for stability of the shoulder joint.

Teres minor and infraspinatus have an important role in abduction of the shoulder where at a position of 90° glenohumeral movement they laterally rotate the humerus to move the greater tubercle away from the coracoacromial arch, allowing continued movement into full elevation.

In addition, supraspinatus acts to resist upward shearing forces of deltoid by initiating abduction. Subscapularis can also assist in adduction of the shoulder.

MUSCLES

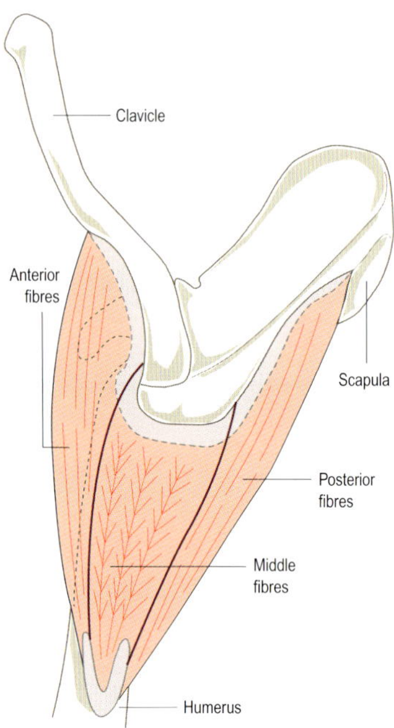

Left deltoid viewed from above

Deltoid – all fibres abduct the arm at the shoulder joint. Anterior fibres flex and medially rotate; posterior fibres extend and laterally rotate the shoulder. *Origin:* anterior fibres – anterior border lateral third of clavicle; middle fibres – outer margin acromion; posterior fibres – lower border spine of scapula. *Insertion:* deltoid tuberosity of humerus. *Nerve supply:* axillary nerve C5, 6.

Middle fibres are extremely powerful because of their multipennate arrangement. With the arm in the anatomical position contraction of deltoid would produce upward shearing of the humerus; contraction of supraspinatus stops this and performs first few degrees of abduction before deltoid is in a position to take over. Middle fibres also work strongly with the shoulder in fixed abduction, allowing function of the arm in raised positions. Damage to the axillary nerve can result in paralysis, leaving only supraspinatus and long head of biceps capable of producing weak abduction at the shoulder.

Pectoralis major – flexes, medially rotates and adducts arm at the shoulder.
Origin: medial half anterior border clavicle; anterior surface sternum; anterior surface upper 6 costal cartilages and 6th rib.
Insertion: via laminated tendon into lateral lip intertubercular groove.
Nerve supply: medial (C8–T1) and lateral (C5, 6, 7) pectoral nerves.

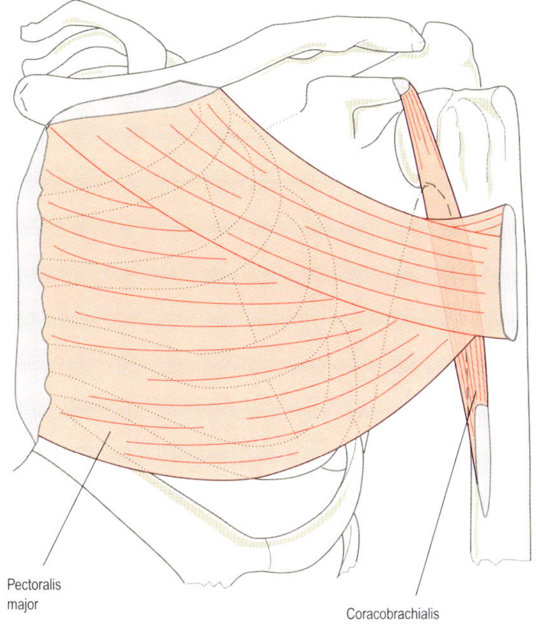

Anterior view left shoulder region

Pectoralis major is a powerful muscle involved in pushing, punching and throwing movements. From full flexion, lower fibres can extend the shoulder to midline against resistance, e.g. pulling body upwards (climbing). With the humerus fixed it can be an accessory muscle of inspiration.

It forms the anterior wall of the axilla where it can be easily palpated.

Coracobrachialis – adducts and flexes arm at the shoulder.
Origin: apex coracoid process (with biceps).
Insertion: medial side shaft of humerus.
Nerve supply: musculocutaneous nerve C6, 7.

MUSCLES

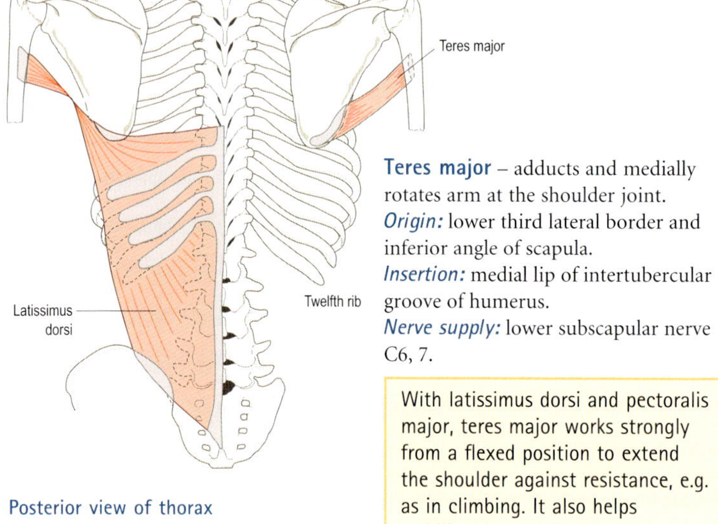

Posterior view of thorax

Teres major – adducts and medially rotates arm at the shoulder joint.
Origin: lower third lateral border and inferior angle of scapula.
Insertion: medial lip of intertubercular groove of humerus.
Nerve supply: lower subscapular nerve C6, 7.

With latissimus dorsi and pectoralis major, teres major works strongly from a flexed position to extend the shoulder against resistance, e.g. as in climbing. It also helps stabilize the shoulder joint.

Latissimus dorsi – adducts, medially rotates and extends the arm at the shoulder joint.
Origin: via thoracolumbar fascia to spinous processes T7–12; all lumbar and sacral vertebrae and their supraspinous and interspinous ligaments; posterior part outer lip iliac crest; lower 4 ribs and inferior angle of scapula.
Insertion: via long flat tendon which twists 180° into floor of intertubercular groove of humerus.
Nerve supply: thoracodorsal nerve C6, 7, 8.

Functionally, latissimus dorsi is a 'climbing' muscle with pectoralis major and teres major. It powerfully extends the shoulder against resistance in activities such as rowing and swimming. If the humerus is fixed, latissimus dorsi can work in 'reverse' and raise the pelvis, e.g. as in walking on crutches. This becomes important in spinal injuries as the level of innervation means it escapes damage in paraplegia and may be the only innervated muscle attaching to the pelvis. Latissimus dorsi also stabilizes the lower end of scapula.

Attachment to the ribs allows latissimus dorsi to assist forced expiration as in coughing.

Latissimus dorsi and teres major form the posterior wall of the axilla where they can be palpated.

MUSCLES

Rotator cuff muscles and shoulder stability

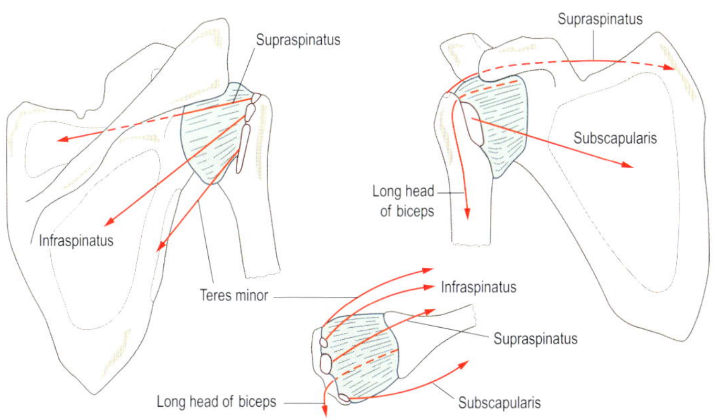

In the diagram above arrows represent direction of pull of the musculotendinous (rotator) cuff and biceps muscles. Attachment of all these muscles is close to the shoulder joint, with four of them blending with the capsule, enhancing their role as stabilizers. Other muscles around the joint also contribute to stability: deltoid, pectoralis major, latissimus dorsi, long head of triceps, coracobrachialis – an essential feature because of the laxity of the capsule and ligaments. This arrangement allows extensive mobility at the shoulder joint whilst muscle tone ensures stability in any position.

The axilla

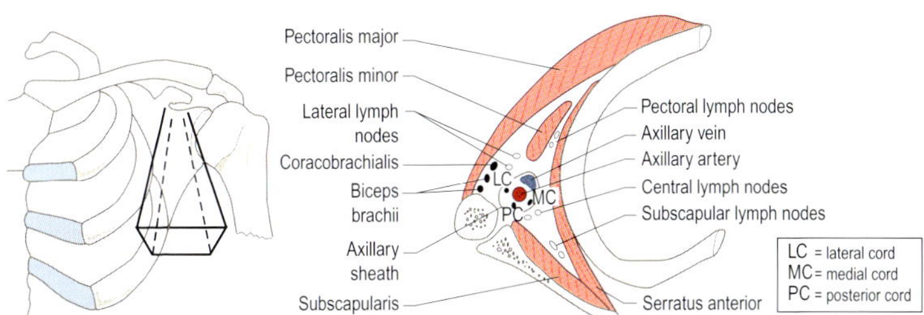

Transverse section

The axilla is a space in the shoulder region. Its boundaries are shown above and it contains a number of important structures, including the cords of the brachial plexus, brachial artery and vein and lymph nodes. Axillary lymph nodes are frequently removed during mastectomy. The cords of the brachial plexus can be damaged in the axilla.

PECTORAL GIRDLE AND SHOULDER JOINT – APPLIED ANATOMY

Combined movements of pectoral girdle and shoulder joint

The pectoral girdle and shoulder joint move together to increase overall range of movement, e.g. abduction through to full elevation combines 60° of lateral rotation of scapula with 120° of abduction of humerus. The humerus will also laterally rotate and the clavicle elevate at its lateral end. Starting in the anatomical position, the first 30° are glenohumeral movement. From this point for every 15°, 10° are glenohumeral and 5° lateral rotation of the scapula (2 : 1 ratio). At 90° glenohumeral movement (actual position 120°) the humerus laterally rotates to allow the greater tuberosity to clear the coroacromial arch, allowing movement to continue to 180°.

If this sequence of movements is altered, e.g. the scapula rotating fully early in the movement, it is often called a 'reversed humeroscapular rhythm'.

Fracture of clavicle

The clavicle is commonly fractured by indirect violence as in a fall on the outstretched arm. The fracture occurs in mid-shaft, partly because of its shape and partly because the strong costoclavicular and coracoclavicular ligaments transfer the sudden pressure to the weaker bone.

Dislocation/subluxation of acromioclavicular joint

This usually occurs as the result of direct violence when the acromion is driven below the clavicle as in a fall on to, or direct blow to, the point of the shoulder.

Dislocation of shoulder

Anterior dislocation is the most common disruption and results from the head of humerus slipping anteriorly out of the glenoid fossa when the arm is forcibly abducted and laterally rotated. This damages the capsule, to which several muscles are attached, destabilizing the joint and leading to recurrence. Surgical intervention may include repair of the capsule and/or shortening of subscapularis tendon.

Subluxation is where supporting rotator cuff muscles fail to maintain normal alignment of joint surfaces, e.g. trauma or paralysis (stroke), making the joint potentially painful if poor handling techniques are applied.

Fracture of surgical neck of humerus

This usually results from a fall on the outstretched arm. The fracture occurs more in the elderly and is often impacted. Displaced fractures are potentially more serious as they may damage vessels or nerves in the axilla. Bone metastases are common here and may lead to pathological fracture.

Supraspinatus

The tendon may become inflamed where it crosses the humeral head below the acromion. Pain is experienced during abduction, usually only between 60 and 120° – the 'painful arc'.

The tendon may also rupture without inflammation, making initiation of abduction difficult.

Frozen shoulder (capsulitis)

This is a painful condition of unknown origin where the soft tissues surrounding the shoulder joint become inflamed and painful, restricting movement in most directions.

ELBOW

BONES

The elbow joint is a synovial hinge joint formed between the distal end of humerus and proximal ends of ulna and head of radius. The *joint line* can be represented by a line drawn between a point 1 cm below the lateral epicondyle and a point 2 cm below the medial epicondyle passing obliquely across the elbow.

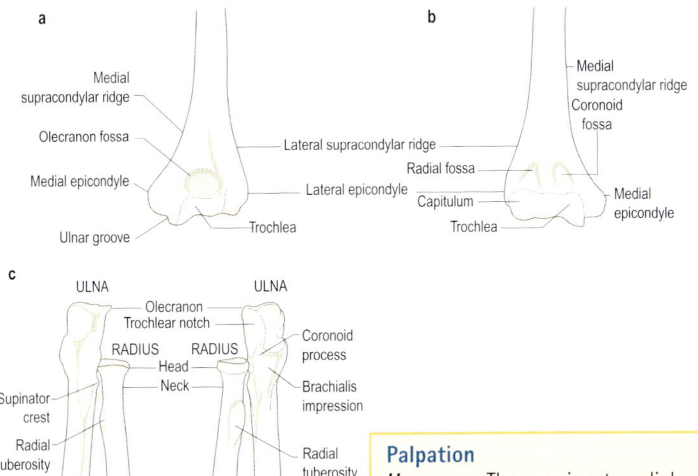

Distal end of humerus (a, anterior; b, posterior) and proximal ends of the radius and ulna (c)

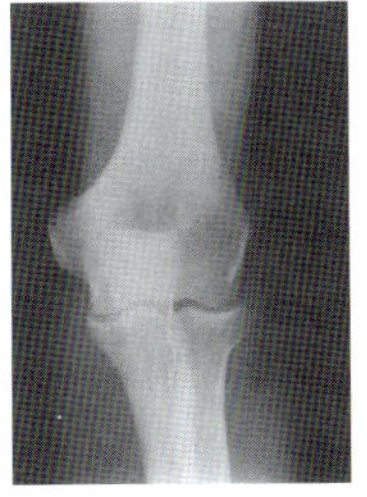

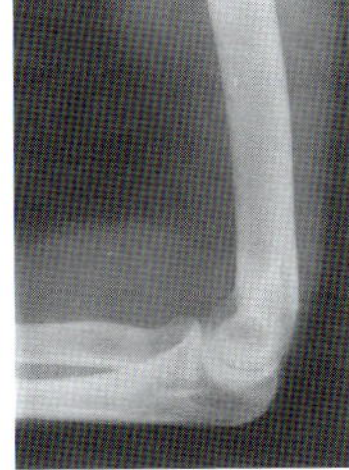

Radiograph of elbow joint: posterior and lateral views

Palpation

Humerus The prominent medial epicondyle can be palpated with the ulnar nerve running in the groove posteriorly. The lateral epicondyle can be palpated at the base of a dimple on the lateral aspect of the elbow. Posteriorly the olecranon fossa can be felt through the relaxed triceps tendon.

Ulna The outline of the olecranon process can be palpated, with the posterior border of the ulna running down from this region.

Radius The head can be palpated on the posterolateral aspect of the elbow, particularly when the elbow is extended: it can be felt rotating during pronation and supination of the forearm.

ARTICULAR SURFACES

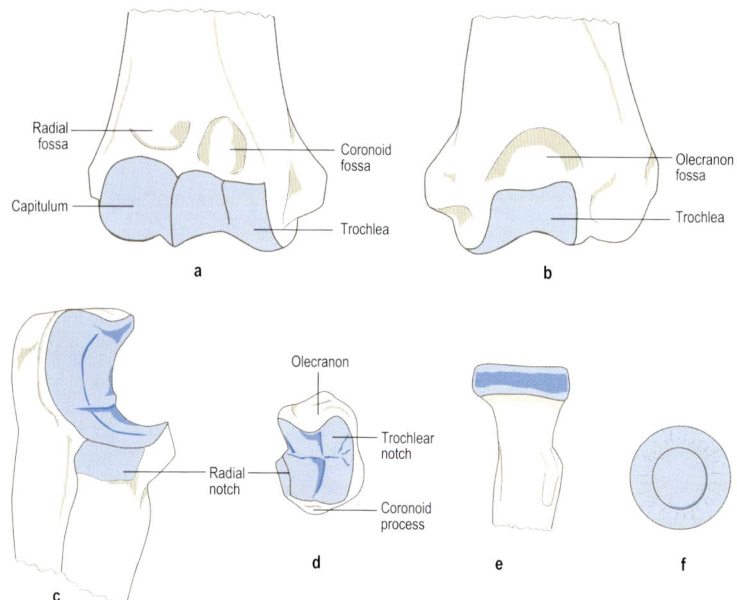

Articular surfaces of the elbow joint: (a) anterior and (b) posterior views of distal end of humerus; (c) lateral and (d) superior views of trochlea of ulna; (e) anterior and (f) superior views of head of radius

Laterally the spherical capitulum of the humerus (a) articulates with the concavity of the head of the radius (e, f).

Medially the trochlea of humerus (a, b) articulates with the trochlear notch of the ulna (c, d).

In *full extension* the proximal part of the olecranon of ulna lies in the olecranon fossa on the posterior aspect of the humerus proximal to the trochlear articular surface.

In *flexion* the coronoid process of ulna and rim of the head of radius lie in the coronoid and radial fossae respectively on the anterior aspect of the humerus proximal to the trochlear surface.

CAPSULE AND SYNOVIAL MEMBRANE

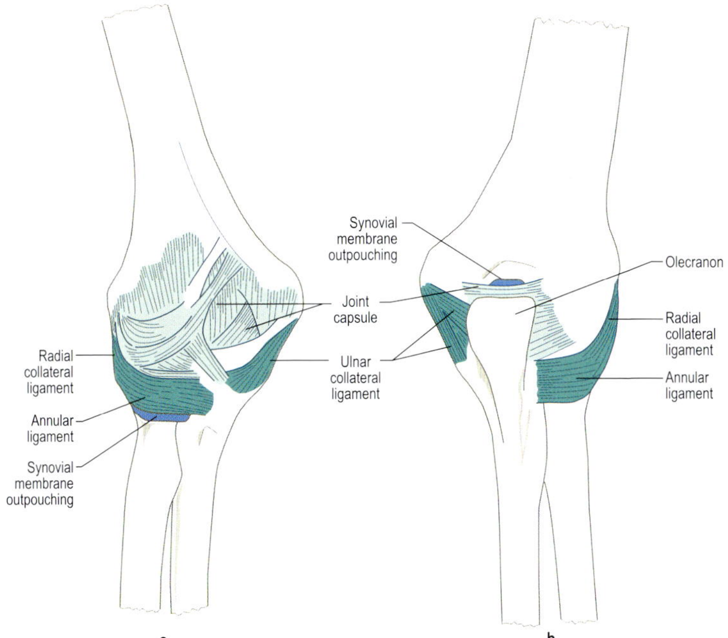

Joint capsule: (a) anterior and (b) posterior aspects

Fibrous capsule Completely surrounds the joint, including the superior radioulnar joint, attaching close to the articular margins and blending medially and laterally with the medial (ulnar) and lateral (radial) collateral ligaments. Anteriorly and posteriorly the deep fibres of brachialis and triceps respectively attach to and reinforce the capsule: these fibres help prevent the capsule becoming trapped during movement.

Synovial membrane Lines the joint capsule attaching to the articular margins of the humerus and ulna. It is reflected onto the humerus to cover the coronoid and radial fossae anteriorly and the olecranon posteriorly. Distally it is prolonged onto the upper part of the deep surface of the annular ligament.

Outpouching of the synovial membrane occurs below the annular ligament and transverse band of the medial collateral ligament and above the transverse capsular fibres across the upper part of the olecranon fossa.

ASSOCIATED LIGAMENTS

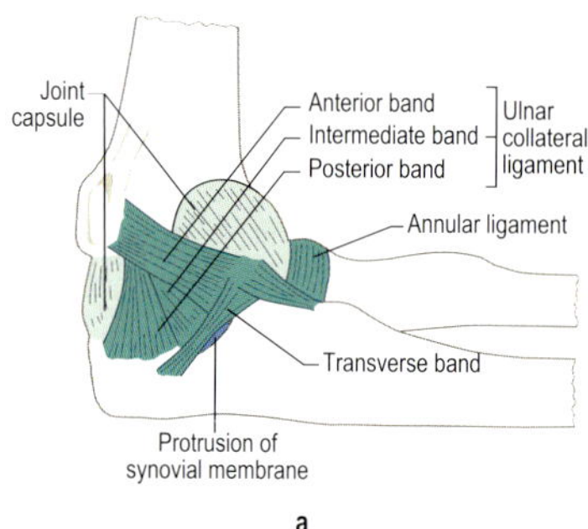

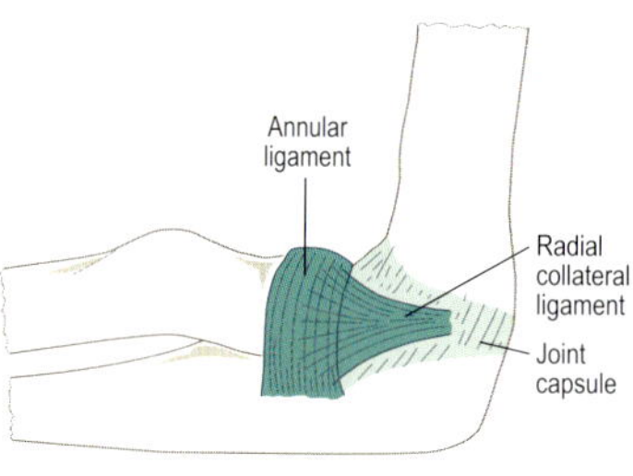

(a) Medial and (b) lateral collateral ligaments

Medial (ulnar) collateral ligament

Runs from the medial epicondyle of humerus fanning out to the medial edge of the coronoid process (anterior band) and medial edge of the olecranon process (posterior band). These two bands are united by a thinner intermediate portion. A thickened transverse band passes between the distal attachments of the anterior and posterior bands.

Lateral (radial) collateral ligament A strong triangular structure passing from the lateral epicondyle to the annular ligament of the superior radioulnar joint: the anterior and posterior margins attach to the margins of the radial notch of ulna.

MOVEMENTS AND STABILITY

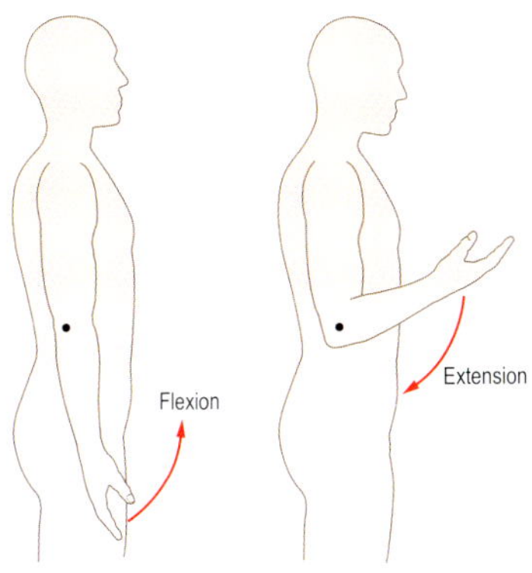

Movements at the elbow joint

Movements Flexion and extension, i.e. a sliding of one articular surface against the other. There is a small amount of abduction/adduction of the ulna associated with pronation and supination of the forearm.

- *Flexion* is movement of the forearm anteriorly: it has an active range of around 145°.
- *Extension* is movement of the forearm posteriorly, i.e. returning the forearm to the anatomical position.

Accessory movements: With the elbow almost fully extended a small degree of abduction and adduction is possible, demonstrated by holding the lower part of the arm steady and applying medial and lateral pressure to the lower end of the forearm.

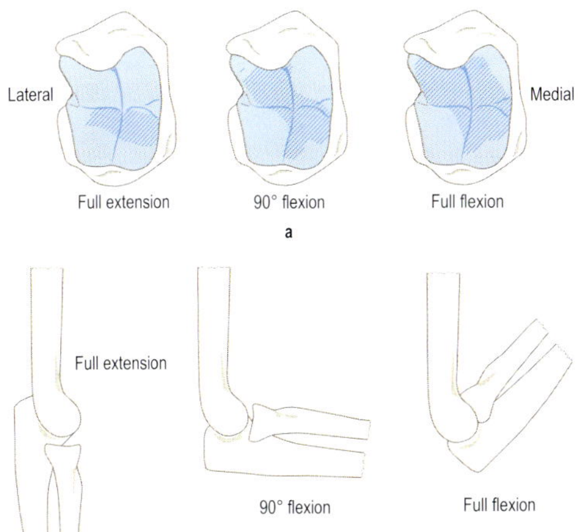

Stability Due to the shape of the articular surfaces and the presence of strong collateral ligaments. During flexion there is an increase in contact area between the trochlea and trochlear notch, with contact being made between the head of the radius and capitulum beyond 90° flexion.

(a) Contact areas of the trochlear notch of ulna; (b) contact of the radial head with the capitulum

MUSCLES

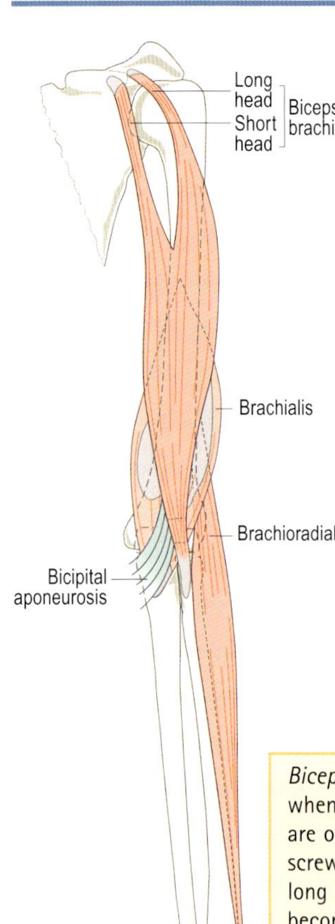

Anterior view left arm

Biceps brachii – flexes elbow and supinates forearm.
Origin: long head – supraglenoid tubercle (scapula), runs through shoulder joint and intertubercular (bicipital) groove; short head – apex of coracoid process of scapula.
Insertion: two bellies fuse and common tendon inserts posterior part of radial tuberosity; bicipital aponeurosis attaches to deep fascia of forearm.
Nerve supply: musculocutaneous nerve C5, 6.

Brachialis – main flexor of elbow.
Origin: distal two-thirds anterior surface of humerus and adjacent septa.
Insertion: inferior part coronoid process and tuberosity of ulna.
Nerve supply: musculocutaneous nerve C5, 6.

Brachioradialis – powerful flexor of elbow with forearm in mid-position.
Origin: upper two-thirds anterior lateral supracondylar ridge of humerus and adjacent septum.
Insertion: lateral surface of radius.
Nerve supply: radial nerve C5, 6.

Biceps brachii has maximum power of supination when elbow is flexed to 90°. Flexion and supination are often performed together, e.g. using screwdriver, corkscrew, turning door handle. The long head can also flex the shoulder and may become inflamed in the bicipital groove. The muscle belly is easily felt at front of upper arm – its tendon is used to test reflex C5/6.

Brachialis lies under biceps brachii and can be felt either side of biceps belly.

Brachioradialis activities with forearm in mid-position include lifting pans, plates, heavy tools (axe, hammer) prior to downswing. Its tendon is easily felt just proximal to the styloid process and is used to test reflex C5/6.

NB: Elbow flexors work to control extension produced by gravity, e.g. eccentric activity (lowering forearm downwards).

MUSCLES

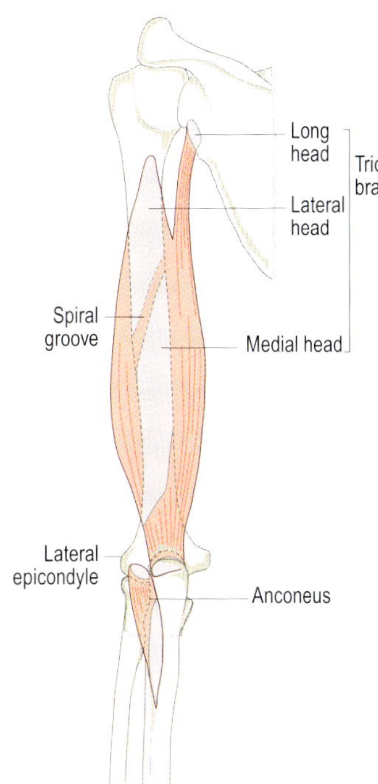

Posterior view left arm

Triceps brachii – main extensor of elbow.

Origin: long head – infraglenoid tubercle of scapula and adjacent glenoid labrum; lateral head – posterior surface of humerus (above and lateral to spiral groove); medial head – posterior surface of humerus (below and medial to spiral groove) and adjacent septa.

Insertion: posterior surface olecranon of ulna and adjacent deep fascia of forearm.

Nerve supply: radial nerve C6, 7, 8.

Anconeus – assists triceps in elbow extension.

Origin: posterior surface lateral epicondyle of humerus.

Insertion: lateral surface of olecranon and posterior surface of ulna.

Nerve supply: radial nerve C7, 8.

Long head of triceps brachii can also adduct and extend arm at shoulder from flexed position. This tendon is easy to see and feel on back of upper arm. The thick tendon of the whole muscle can be felt just above olecranon process and is used to test reflex C7/8.

Anconeus can also alter the axis of pronation/supination of the forearm by moving the ulna.

Once the elbow is flexed, gravity may provide force for extension with *flexors* working eccentrically to control movement. Triceps works in this position if speed/strength is needed, e.g. chopping wood, using hammer. *Extensors* also work against resistance in activities such as pushing up out of chair, lowering down into chair, propelling wheelchair, using crutches, etc.

RELATIONS

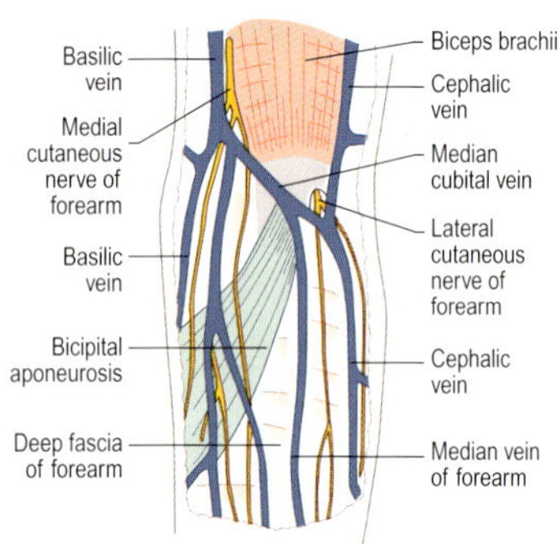

Superficial structures (anterior view)

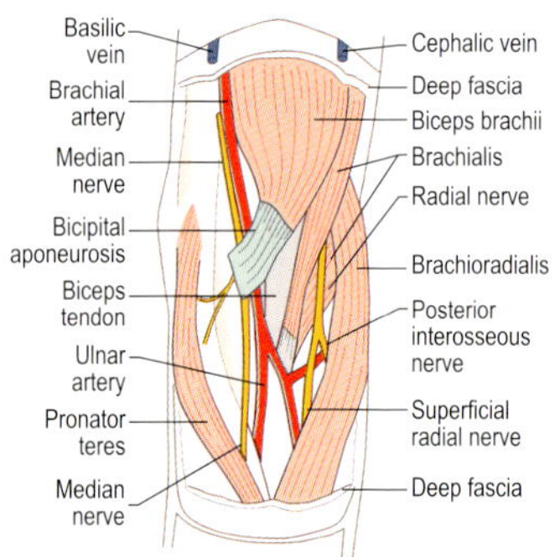

Deep structures (anterior view)

Cubital fossa The cubital fossa is a triangular space bounded superiorly by an imaginary line between the medial and lateral epicondyles and by the converging medial borders of pronator teres medially and brachioradialis laterally. The floor is formed by brachialis, with supinator inferomedially.

Deep fascia of the forearm covers the fossa, reinforced medially by the bicipital aponeurosis, separating superficial veins (cephalic, basilic, median cubital) and nerves from deeper structures.

Passing through the fossa are (from medial to lateral): *median nerve*, which enters the anterior forearm through pronator teres; *brachial artery*, dividing into the radial and ulnar arteries; *tendon of biceps brachii* inserting onto the radial tuberosity; *radial nerve* dividing into deep (posterior interosseous) and superficial branches – deep branch enters the posterior forearm through supinator. The *ulnar nerve* passes posterior to the medial epicondyle, entering the forearm through flexor carpi ulnaris.

Large superficial veins, e.g. median cubital, frequently used for venepuncture.

Brachial artery is a common site for determining blood pressure.

APPLIED ANATOMY

Fractures

Supracondylar fractures are common in children falling on an outstretched arm. The fracture is often comminuted and may be complex to treat with vascular (brachial artery) and nerve (median) complications. The *olecranon* may be fractured in direct falls onto the point of the elbow, the humerus bearing down into the olecranon at its narrowest point. *Avulsion fracture of the olecranon* process may also result from a sudden powerful pull of the triceps attachment.

Dislocation

The elbow can be dislocated by a fall on the outstretched arm. The lower end of humerus may displace anteriorly over the coronoid process, which may also be fractured.

Myositis ossificans (or ectopic ossification)

Trauma or fracture close to the elbow joint can trigger unwanted bone growth around the joint, forming a solid mass in the adjacent soft tissues which is easily seen on X-ray.

Ulnar nerve entrapment

Compression of the ulnar nerve can occur where it runs through its fibrous tunnel behind the medial epicondyle. It may result from an abnormal valgus angle caused by earlier fracture or by joint deformity/damage due to rheumatoid arthritis or osteoarthritis.

Tennis elbow

Tenderness and pain caused by a tear in or near the insertion of the common extensor tendon on the lateral condyle of humerus. It can result from sudden flexion of the wrist while the extensors are contracted, e.g. awkward backhand tennis stroke, but is more often the result of repeated gripping in everyday activities, so-called cumulative trauma disorders.

Golfer's elbow

Similar to tennis elbow but affects the common flexor tendon where it attaches to the medial condyle of humerus. Again repetitive strain of the flexor origin will produce pain and tenderness in the region.

Olecranon bursitis

Inflammation of the olecranon bursa near the point of the elbow may be caused by trauma or infection. It is also associated with inflammatory conditions such as rheumatoid arthritis and gout.

FOREARM

BONES

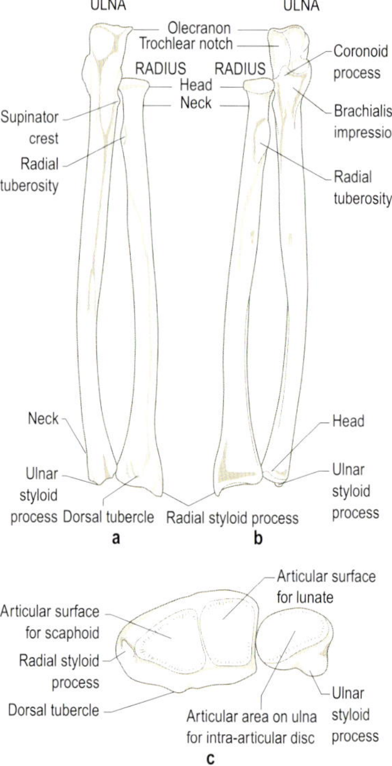

(a) Posterior and (b) Anterior aspects of radius and ulna; (c) distal articular surfaces

The shorter radius (lateral) and longer ulna (medial) articulate proximally with the humerus at the elbow joint and contribute to the wrist distally. Their shafts are joined by an interosseous membrane: synovial pivot joints at each end permit pronation and supination.

Radius Rounded proximal head with concave superior and outer flattened surfaces leading to constricted neck and shaft. Radial tuberosity lies anteromedial on upper shaft. Expanded distal end with five distinct surfaces: lateral extending on to styloid process; concave medial forming the ulnar notch; convex posteriorly, grooved by tendons with the prominent dorsal (Lister's) tubercle in the middle; smooth anterior; concave distal articular divided into lateral triangular and medial quadrilateral areas.
Ulna Expanded proximal end with posterior olecranon and anterior coronoid processes surrounding the trochlear notch. Radial notch lies on the lateral side of the coronoid process, below which is the supinator fossa and crest. The shaft has three borders and three surfaces. Distally the shaft narrows then expands to the small rounded head with the downward-projecting styloid process posteromedially.

Palpation

Radius The head can be palpated on the posterolateral aspect of the elbow and felt rotating against the capitulum during pronation/supination. The lateral side of the distal shaft and styloid process can be palpated, as can the dorsal tubercle posteriorly proximal to the wrist.

Ulna The olecranon process ('point' of the elbow) can be seen and felt posteriorly, together with the whole of the posterior border. Distally the neck, head and styloid process can be palpated. In full pronation the round head stands out from the back of the wrist.

ARTICULAR SURFACES

Synovial pivot joints

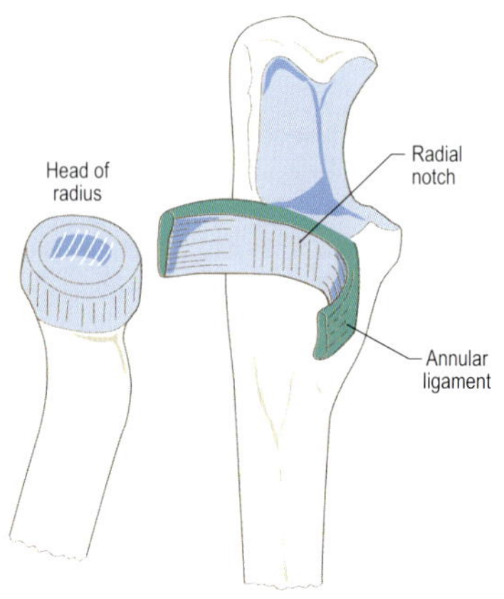

Superior radioulnar joint (opened)

Superior radioulnar joint The articulation is between the bevelled circumference of the head of the radius within the fibro-osseous ring formed by the radial notch of the ulna and annular ligament.

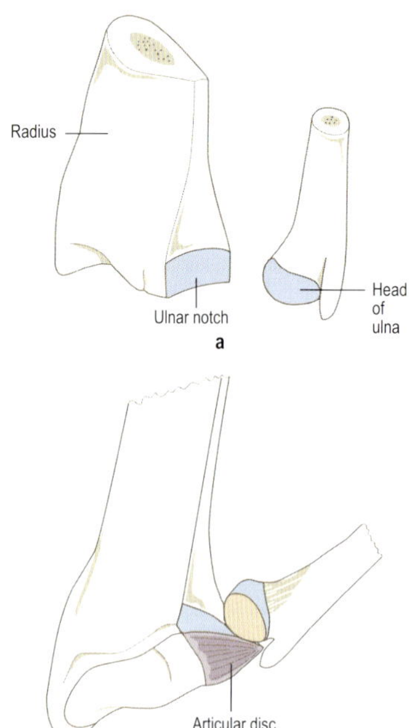

Inferior radioulnar joint The articulation is between the head of the ulna and the ulnar notch of the radius. The distal end of the head articulates with a triangular fibrocartilaginous intra-articular disc attached to the lateral side of the root of the ulna styloid process and inferior margin of the ulnar notch of radius.

Inferior radioulnar joint: (a) opened; (b) showing intra-articular disc

CAPSULE AND SYNOVIAL MEMBRANE

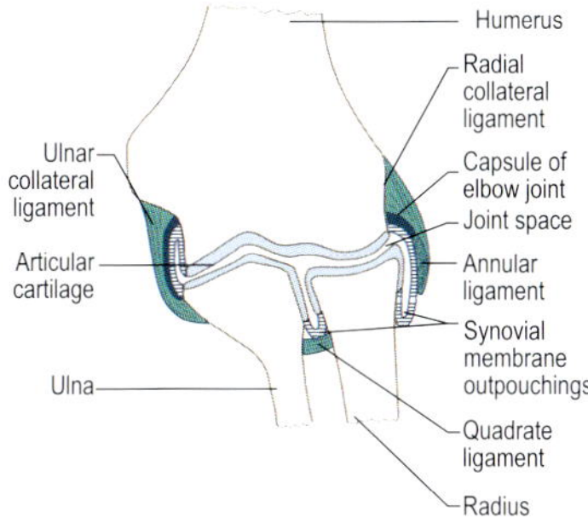

Coronal section through superior radioulnar and elbow joints

Superior radioulnar joint The superior radioulnar joint is continuous with the elbow joint and shares the same fibrous joint capsule attaching to the superior border and outer surface of the annular ligament.

The *synovial membrane* attaches to the upper margin of the annular ligament: from the lower margin of the annular ligament it extends below as a redundant fold loosely attached to the neck of radius.

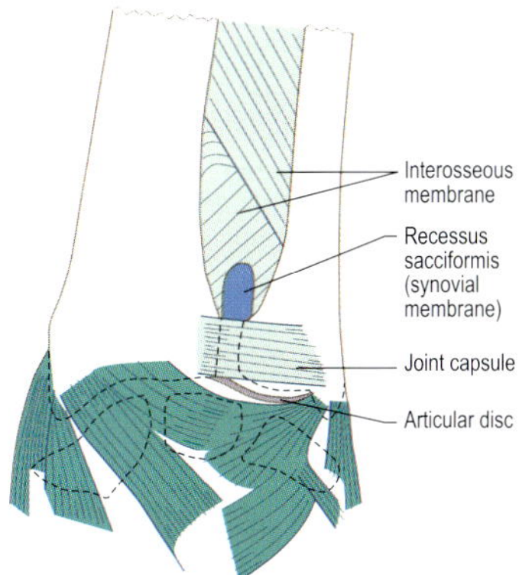

Inferior radioulnar joint showing joint capsule and outpouching of synovial membrane (recessus sacciformis)

Inferior radioulnar joint The weak and loose fibrous capsule is formed by transverse bands of fibres attached to the anterior and posterior margins of the ulnar notch of radius and to corresponding areas on the ulna: inferiorly the bands blend with the anterior and posterior edges of the disc but superiorly they remain separate.

The *synovial membrane* extends above the joint capsule between the radius and ulna anterior to the interosseous membrane (recessus sacciformis).

LIGAMENTS AND INTEROSSEOUS MEMBRANE

Annular ligament Flexible fibrocartilaginous band attached to anterior and posterior margins of radial notch of ulna: it forms four-fifths of the articular surface of superior radioulnar joint. The narrower inferior diameter cups the radial head, preventing its displacement. The upper part of the ligament is lined with fibrocartilage continuous with the hyaline cartilage of the radial notch. Superiorly it fuses with the lateral collateral ligament of the elbow and blends with the joint capsule anteriorly and posteriorly.

Quadrate ligament From the lower border of the radial notch of the ulna to the adjacent medial surface of the radial neck proximal to the radial tuberosity.

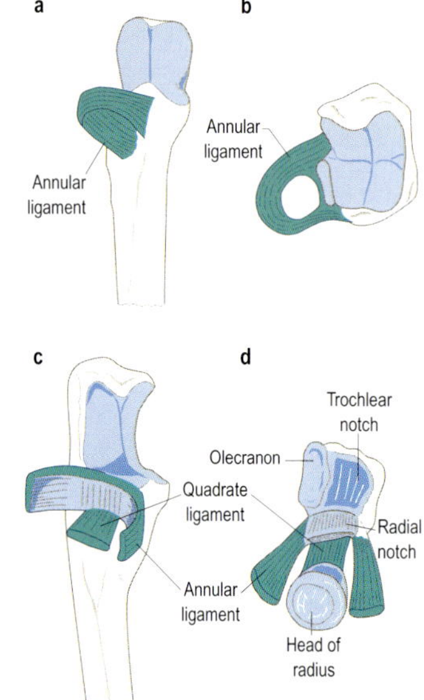

Annular (a–d) and quadrate (c, d) ligaments

Its fibres run in such directions between the bones that the overall tension within the ligament is constant in all positions of pronation/supination.

Interosseous membrane Between the interosseous borders of the radius and ulna, with the fibres passing predominantly inferomedially: these serve to transmit forces from the radius to the ulna and hold the bones together. Deficient superiorly, but continuous with fascia over the posterior surface of pronator quadratus. Openings superiorly and inferiorly transmit the posterior and anterior interosseous vessels respectively.

Provides attachment for the deep flexors and extensors of the forearm.

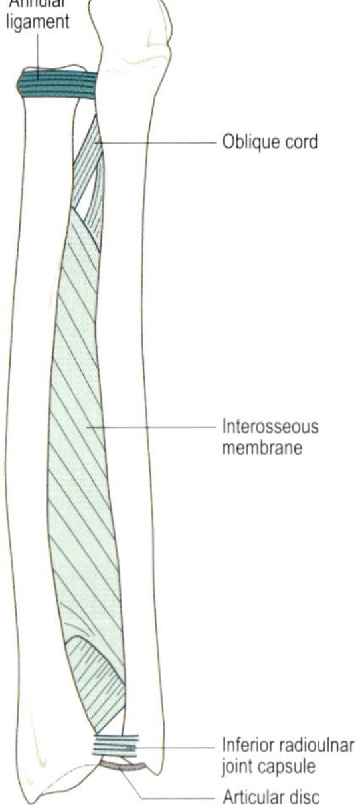

Interosseous membrane

MOVEMENTS

Pronation and *supination* of the forearm occur at the radioulnar joints.

From the anatomical position pronation is turning the palm to face posteriorly with the thumb then lying medially: it is produced by pronators teres and quadratus. The reverse movement is supination: it is produced by supinator and biceps, of which biceps is the more powerful.

At the *superior radioulnar joint* the head of radius rotates within the fibro-osseous ring. There are three other components of the movement:

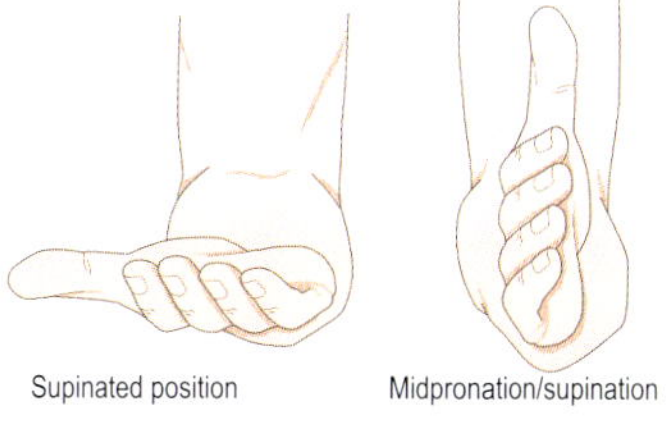

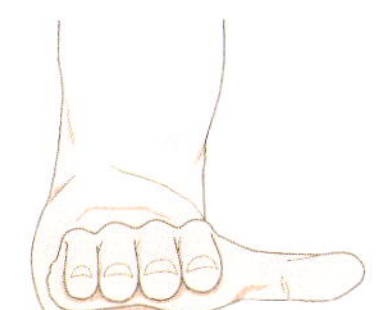

Pronation and supination

1. rotation of the superior surface of the head against the capitulum
2. lateral displacement of the radial head
3. tilting of the plane of the radial head inferolaterally.

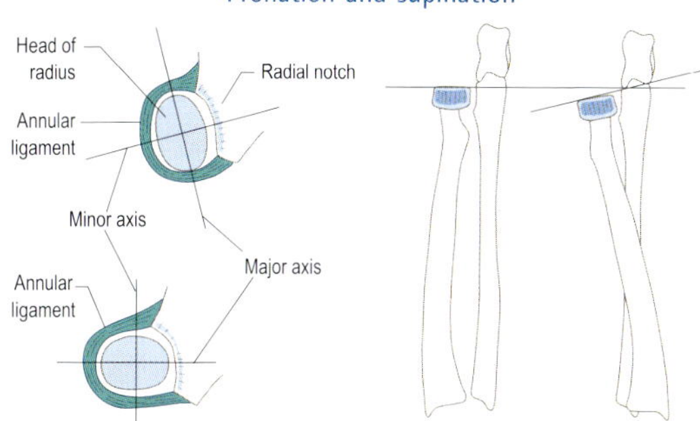

Movement of radius during pronation/supination

At the *inferior radioulnar joint* the lower end of the radius rotates around the head of the ulna, which at the same time undergoes extension and lateral displacement. These latter movements are small and do not involve any rotation of the ulna.

Accessory movements: Anteroposterior accessory movements of the head of radius can be elicited at the superior joint and of the head of ulna at the inferior joint by gripping the head and moving it with respect to the ulna and radius respectively.

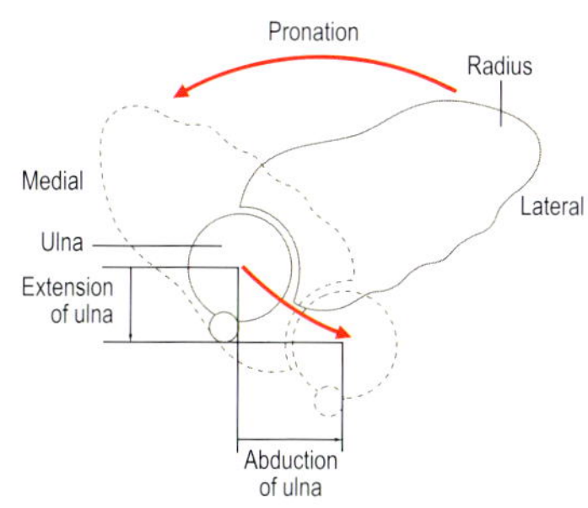

Movement of radius and ulna during pronation

MUSCLES

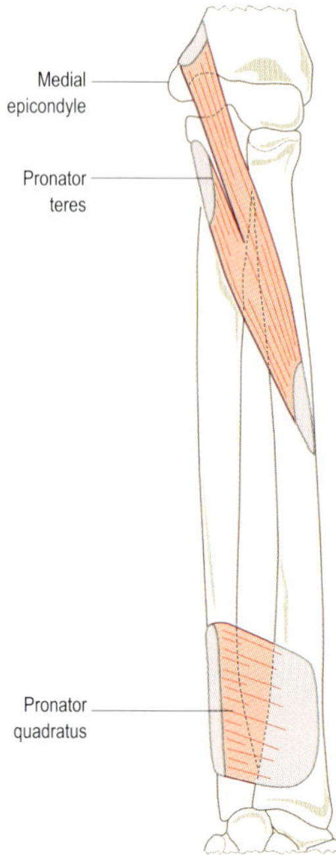

Anterior view left forearm

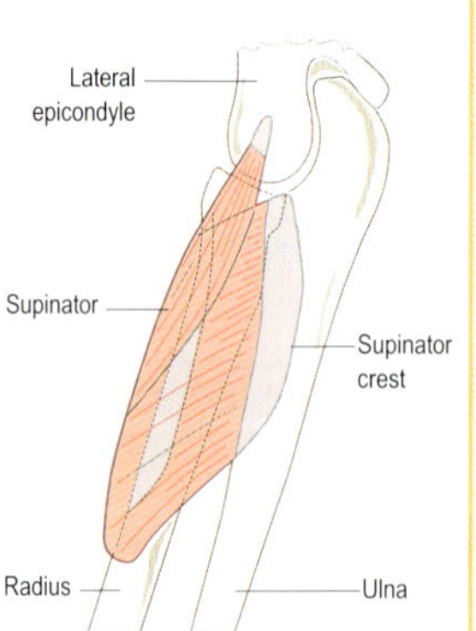

Lateral view left arm

Pronator teres – pronates forearm and is weak flexor of elbow.
Origin: common flexor origin, medial epicondyle of humerus; pronator ridge of ulna.
Insertion: middle lateral surface of radius.
Nerve supply: median nerve C6, 7.

Pronator quadratus – pronator of forearm.
Origin: lower quarter anterior surface of ulna.
Insertion: lower quarter anterior surface of radius.
Nerve supply: anterior interosseous branch, median nerve C8, T1.

Supinator – supinates forearm in any position.
Origin: lateral epicondyle of humerus, radial collateral ligament, annular ligament, supinator crest of ulna.
Insertion: posterior, lateral and anterior aspects upper third of radius.
Nerve supply: posterior interosseous branch of radial nerve C5, 6.

Pronation is used functionally in activities such as removing a screw using a screwdriver in the right hand. Supination is a more powerful movement being produced by supinator and biceps brachii and is used in functional activities such as putting in a screw or a corkscrew with the right hand.

Anconeus can abduct the ulna to move the distal axis of pronation/supination so that it allows a 'centred' rotation as in keeping the tip of a screwdriver in the notch of a screw.

Pronator quadratus holds the lower ends of radius and ulna together, resisting upward pressure when weight-bearing through the hand.

Palpation

Pronator teres can be felt at the medial border of the cubital fossa during resisted pronation. Supinator can be felt over the posterior part upper third of radius during resisted supination.

WRIST AND CARPUS

BONES

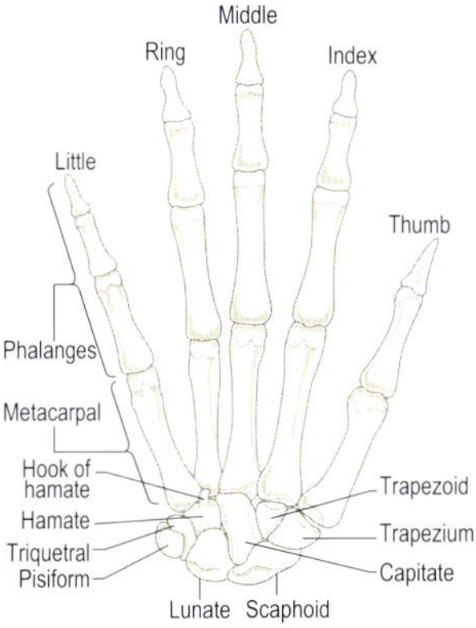

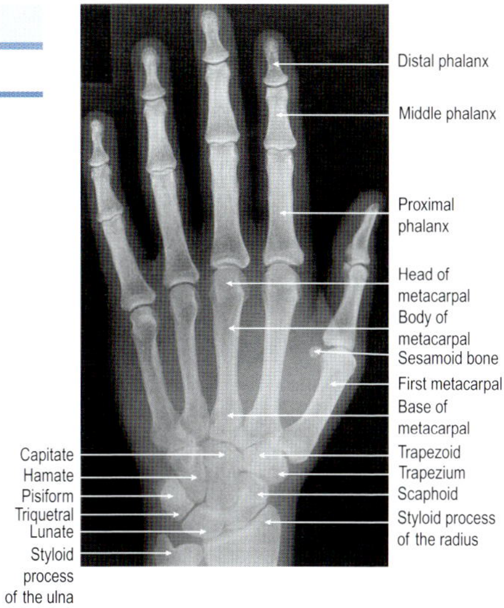

Carpal bones

The wrist consists of eight individual small bones (the carpus) arranged around the capitate, but usually considered as being in two rows (proximal and distal). In the proximal row from lateral to medial are the scaphoid, lunate, triquetral and pisiform, and in the distal row the trapezium, trapezoid, capitate and hamate.

The proximal row, except the pisiform, articulates with the distal end of radius and the intra-articular disc forming the *radiocarpal joint*. The articulation between proximal and distal rows forms the *midcarpal joint*. The distal row articulates with the metacarpal bases forming the *carpometacarpal joints*.

Palpation

Radius Proximal to the wrist the dorsal tubercle can be palpated posteriorly, with the styloid process laterally between the extensor tendons of the thumb.

Ulna The neck, head and styloid process can be palpated with the forearm pronated.

Carpus Laterally, the tubercles of scaphoid and trapezium can be palpated. Medially, the pisiform can be palpated as can the hook of the hamate distal to this, deep to the hypothenar muscles.

ARTICULAR SURFACES

Radiocarpal joint A synovial ellipsoid joint. The distal surface of radius and intra-articular disc form a continuous concave ellipsoid articular surface. The proximal row of carpal bones, united by interosseous ligaments, forms a continuous convex articular surface. Plane synovial joints exist between adjacent surfaces of scaphoid, lunate and triquetral.

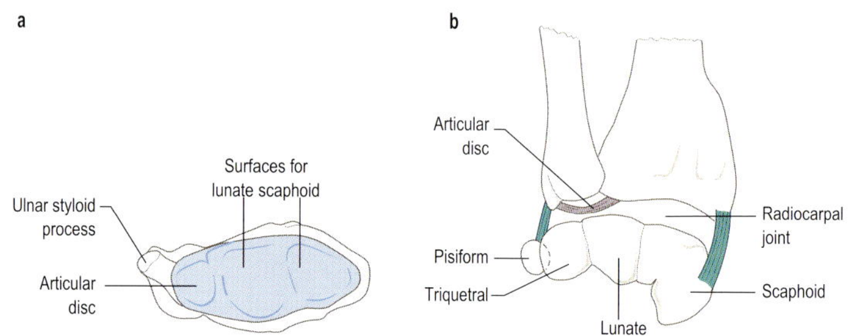

Radiocarpal joint: (a) distal surface of radius and articular disc; (b) posterior aspect

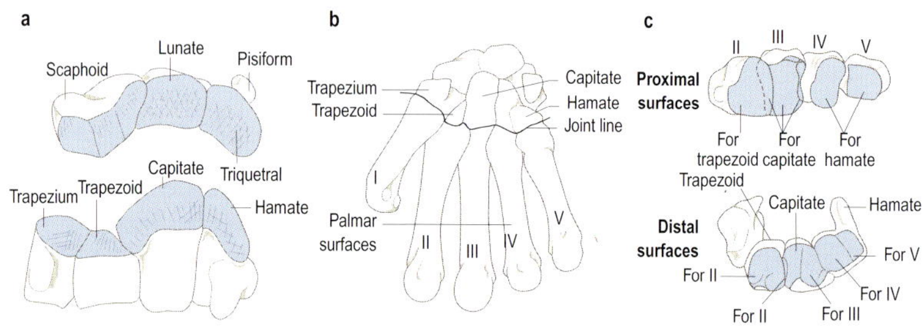

Midcarpal (a, c) and carpometacarpal (b, c) joints

Midcarpal joint Between proximal and distal rows of carpal bones: the distal surface of scaphoid is slightly convex, while the surfaces of lunate and triquetral are concave in all directions.

Laterally the trapezium and trapezoid articulate with the scaphoid; centrally the head of capitate articulates with the scaphoid and lunate; medially the hamate articulates with lunate and triquetral.

Common carpometacarpal (CMC) joints Plane synovial joints between distal surfaces of trapezoid, capitate and hamate and the bases of the medial four metacarpals. The articular surfaces are flat, except for the slightly bevelled joint surfaces between the hamate and 5th metacarpal.

CAPSULE AND SYNOVIAL MEMBRANE

Separate fibrous capsules completely surround the radiocarpal, midcarpal and common carpometacarpal (CMC) joints. There is usually direct communication between all of the above joint spaces.

The *radiocarpal* joint capsule attaches proximally to the distal margins of radius and ulna anteriorly and posteriorly, and to the radial and ulnar styloid processes laterally and medially: distally it attaches to the anterior and posterior margins of the proximal row of carpals.

The *midcarpal* joint capsule consists of irregular bands of fibres (palmar and dorsal intercarpal ligaments) between the rows of bones.

The *CMC* joint capsule extends between the distal carpal bones and bases of the four medial metacarpals. Distally the joint space extends between the metacarpal bases.

Synovial membrane lines all non-articular surfaces.

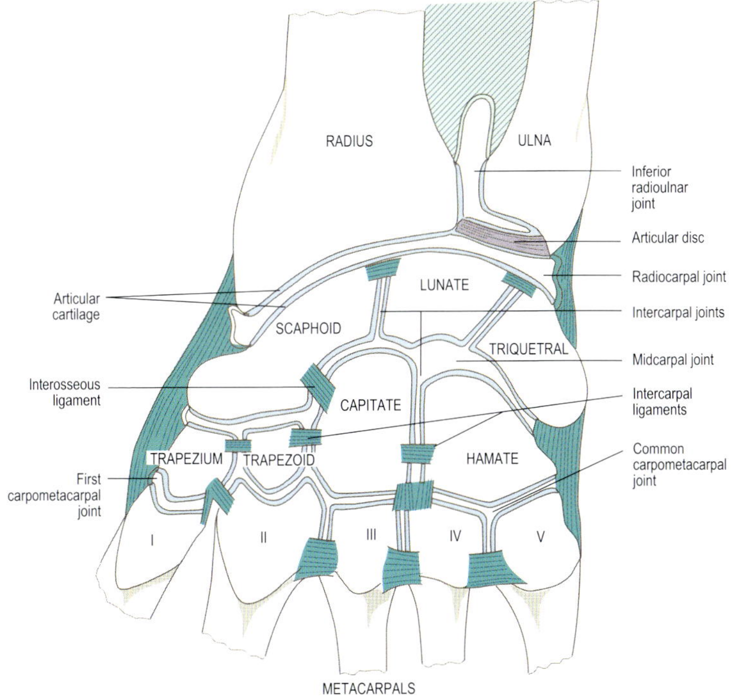

Coronal section through wrist showing capsule and intercarpal ligaments

LIGAMENTS

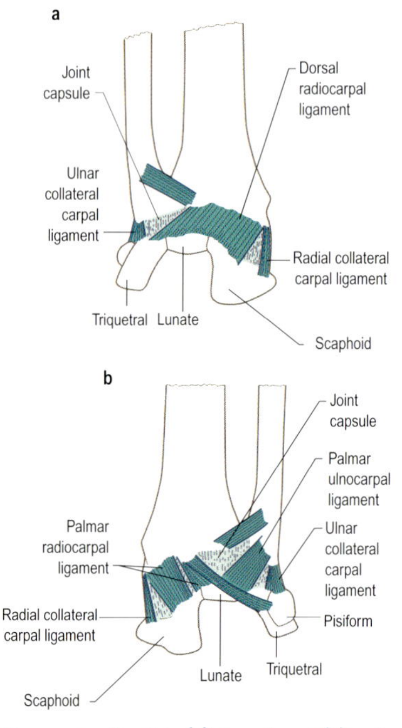

Ligaments of wrist: (a) dorsal and (b) palmar aspects

Radiocarpal joint The *radial* and *ulnar collateral carpal ligaments* reinforce and strengthen the lateral and medial aspects of the joint capsule. The radial ligament passes from the tip of the radial styloid process to the scaphoid, while the ulnar ligament passes from the ulnar styloid process to pisiform and triquetral. Anteriorly and posteriorly are the *palmar* and *dorsal radiocarpal ligaments* passing from the radius to scaphoid, lunate and triquetral.

Midcarpal joint At the sides the joint capsule is strengthened by *radial* and *ulnar collateral ligaments* passing between scaphoid and trapezium and the triquetral and hamate respectively. Anteriorly the *palmar intercarpal (radiate capitate) ligament* runs between the capitate and proximal row, while posteriorly the *dorsal intercarpal ligament* runs between the two rows of bones.

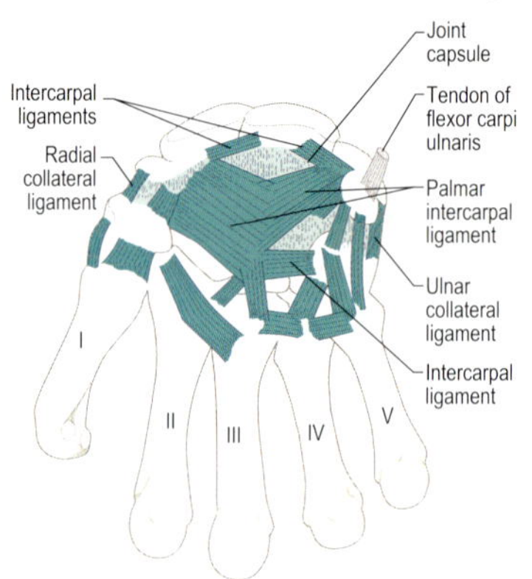

Arrangement of ligaments on palmar aspect of wrist

Common CMC joint The *palmar* and *dorsal carpometacarpal ligaments* are thickenings of the joint capsule passing between the two rows of carpal bones.

Interosseous ligaments pass between adjacent carpal bones.

MOVEMENTS

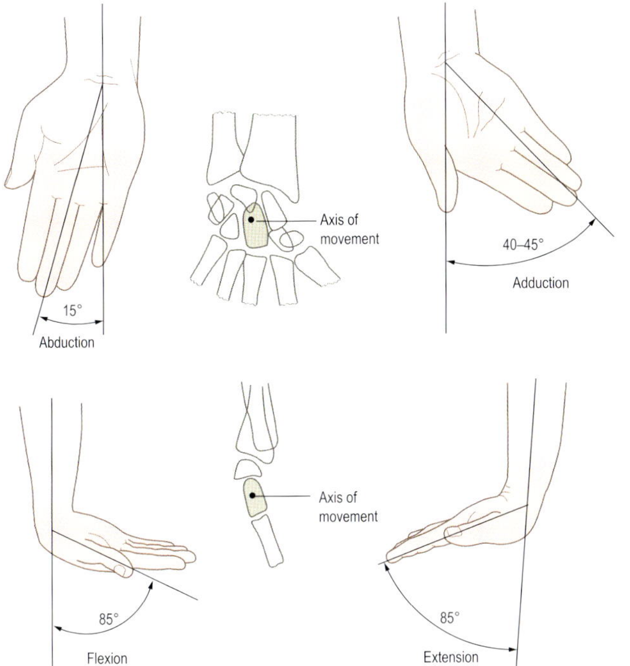

Movements at the wrist

Movements at the wrist are *flexion*, *extension*, *abduction* and *adduction*, each occurring simultaneously at the radiocarpal and midcarpal joints. Flexion and extension each has a range of 85°, abduction 15° and adduction 45°. Flexion is limited by tension in the extensor tendons, being greatly reduced if the fingers are fully flexed. Abduction is limited by the radial styloid process coming into contact with the scaphoid and apposition of the lateral articular surfaces of the midcarpal joint.

Flexion and extension occur about a single transverse axis and abduction and adduction about a single anteroposterior axis: both axes pass through the head of capitate.

Small movements occur at the intercarpal joints accompanying and facilitating movements at the radiocarpal and midcarpal joints.

Accessory movements: Anteroposterior gliding of the proximal row of carpal bones against the radius and articular disc can be produced by an appropriate force, as can a longitudinal movement. Similar movements can be produced between the proximal and distal rows of carpal bones.

MUSCLES

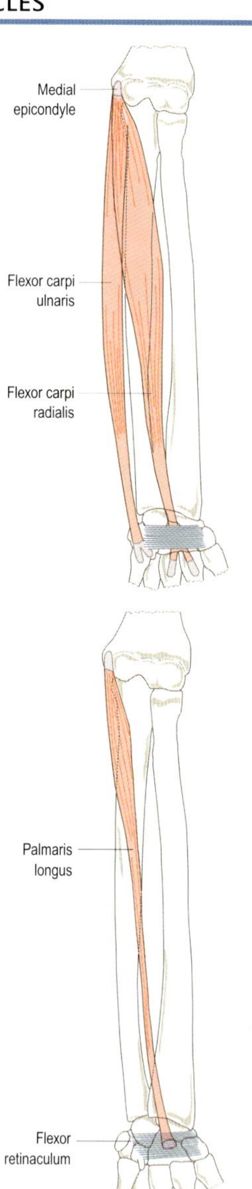

Anterior views of left forearm and wrist

Flexor carpi ulnaris – produces flexion and ulnar deviation (adduction) of wrist.
Origin: medial epicondyle of humerus via common flexor tendon; medial border olecranon; upper two-thirds posterior border ulna via an aponeurosis.
Insertion: pisiform then via ligaments to hook of hamate and 5th metacarpal.
Nerve supply: ulnar nerve C7, 8.

Flexor carpi radialis – produces flexion and radial deviation (abduction) of wrist.
Origin: medial epicondyle of humerus via common flexor tendon.
Insertion: palmar surface bases of 2nd and 3rd metacarpals.
Nerve supply: median nerve C6, 7.

Palmaris longus – a vestigial muscle lying between flexors carpi ulnaris and radialis, running from the medial epicondyle to the flexor retinaculum. It is a weak flexor of the wrist supplied by the median nerve, C8.

> The tendon of palmaris longus is often 'harvested' for tendon transplant/repair.
>
> Radial and ulnar deviation are normally produced by combined action of the flexors and extensors carpi.
>
> The tendon of flexor carpi ulnaris can easily be felt proximal to the pisiform. If palmaris longus is present the tendon runs centrally across the wrist, blending with the flexor retinaculum and palmar aponeurosis. The tendon of flexor carpi radialis lies on its lateral side.

MUSCLES

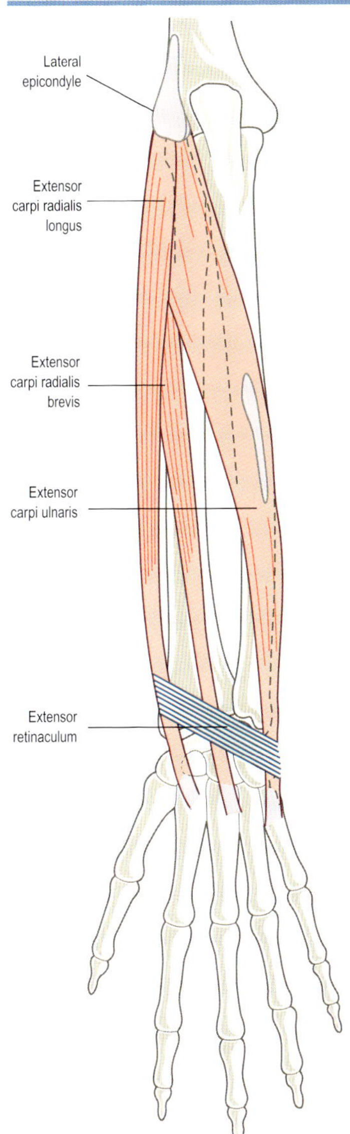

Posterior view left forearm and wrist

Extensor carpi radialis longus – extends and radially deviates (abducts) wrist.
Origin: lower third lateral supracondylar ridge of humerus and adjacent intermuscular septum.
Insertion: posterior surface base of 2nd metacarpal.
Nerve supply: radial nerve C6, 7.

Extensor carpi radialis brevis – extends and radially deviates (abducts) wrist.
Origin: lateral epicondyle of humerus via common extensor tendon.
Insertion: posterior surface base of 3rd metacarpal.
Nerve supply: posterior interosseous branch of radial nerve C6, 7.

Extensor carpi ulnaris – produces extension and ulnar deviation of wrist.
Origin: lateral epicondyle of humerus via common extensor tendon; via aponeurosis to posterior border of ulna.
Insertion: medial side base of 5th metacarpal.

All three tendons pass deep to the extensor retinaculum, enclosed within synovial sheaths.

> Functionally these three extensors work strongly during gripping. The wrist must be in extension for a strong grip as this position allows the 'slack' to be taken up in the finger flexors (active insufficiency) – it is impossible to grip strongly with the wrist flexed. They also work as synergists during gripping, preventing unwanted continued action of the finger flexors which would flex the wrist. This explains why paralysis of these muscles following a radial nerve lesion has a severe effect on function of the hand.

RELATIONS

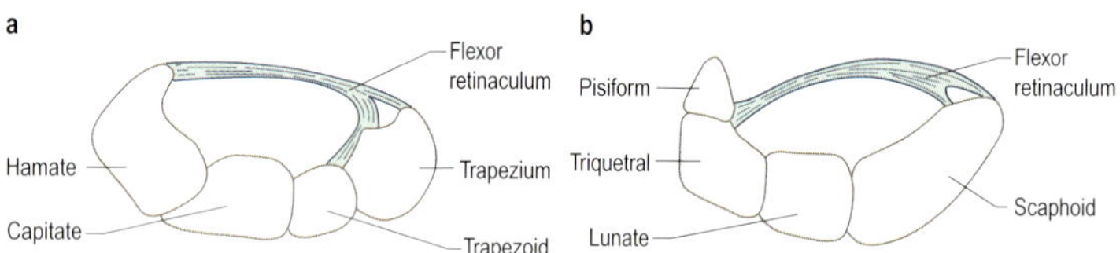

Attachment of flexor retinaculum to carpus: (a) proximal row; (b) distal row

Illustrated above, the flexor retinaculum is shown at the levels of proximal and distal rows of the carpus. The fibro-osseous structure formed by the carpal bones posteriorly and the fibrous flexor retinaculum anteriorly is known as the carpal tunnel. The resulting inflexible space prevents structures passing through it from 'bowstringing' during flexion of the wrist but makes the carpal tunnel an anatomical space vulnerable to increases in pressure, e.g. carpal tunnel syndrome.

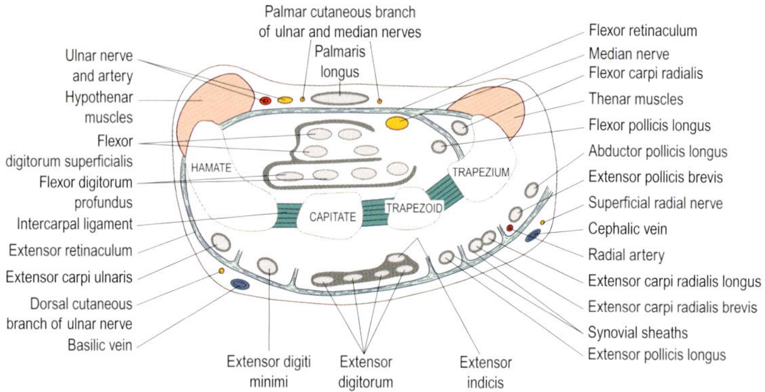

Structures around the wrist (transverse section)

This transverse section shows the position of all major nerves, blood vessels and tendons crossing the wrist and their relationship to the flexor and extensor retinacula. It also demonstrates where synovial sheaths are shared.

RELATIONS

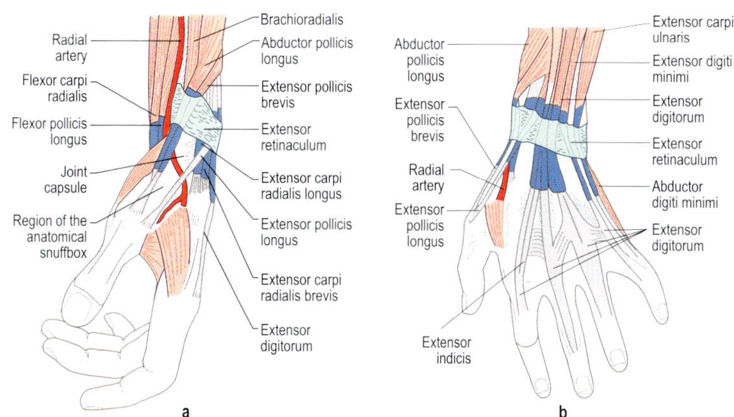

(a) Lateral and (b) posterior relations at the wrist

The extensor tendons enter the hand posterior to the carpal bones deep to the extensor retinaculum, which attaches to the distal part of the anterior surface of the radius laterally and the distal end of the ulna, pisiform and triquetral medially. Fibrous septa from the deep surface of the retinaculum pass to ridges on the radius and ulna creating six compartments.

1. Over the lateral surface of the radial styloid process, scaphoid and trapezium pass the tendons of abductor pollicis longus and extensor pollicis brevis in the same compartment.
2. In the adjacent compartment over the radius lateral to the dorsal tubercle over the scaphoid and medial part of the trapezium are the tendons of extensors carpi radialis longus and brevis.
3. In a groove on the medial side of the dorsal tubercle passes the tendon of extensor pollicis longus, using the tubercle as a pulley so that it deviates laterally towards the thumb as it passes over the scaphoid and trapezium.
4. Over the most medial part of the dorsum of the radius and then adjacent parts of the scaphoid, lunate and capitate are the tendons of extensor digitorum with that of extensor indicis: all five tendons share the same synovial sheath.
5. Over the posterior surface of the inferior radioulnar joint, lunate and adjacent surfaces of the capitate and hamate passes the tendon of extensor digiti minimi.
6. In the groove on the back of the ulna and then the triquetral passes the tendon of extensor carpi ulnaris.

The *radial artery* enters the hand by passing through the anatomical snuffbox over the scaphoid and trapezium deep to the tendons of abductor pollicis longus and extensors pollicis longus and brevis, and then between the two heads of the 1st dorsal interosseous.

HAND

BONES

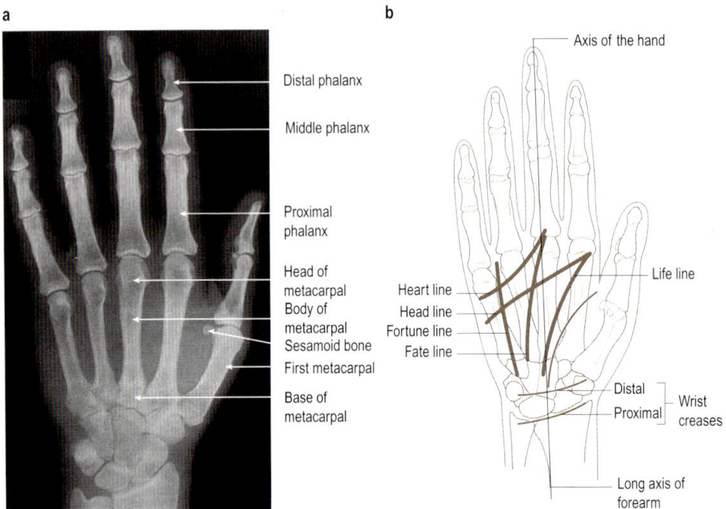

(a) Radiograph of hand; (b) relation of 'lines' to bones of wrist and hand

The *metacarpals*, numbered M1 to M5 from lateral to medial, are long bones each with a quadrilateral base, shaft and a distal rounded head. The metacarpal bases articulate with the distal row of carpal bones: M1 with the trapezium; M2 the trapezium, trapezoid and capitate; M3 the capitate; M4 and M5 the hamate.

There are 14 *phalanges* in each hand, three in each finger and two in the thumb. Each has a large proximal end, a curved shaft and a rounded pulley-shaped distal head. The heads of the distal phalanx are expanded to support the pulp pad of the digits.

By convention the digits are described by name rather than number, from lateral to medial: thumb; index, middle, ring and little fingers.

Palpation

Metacarpals With fingers flexed the heads of the metacarpals can be palpated as the knuckles. On the dorsal surface the shaft is easily felt, as is the line of the carpometacarpal joint.

Phalanges The heads of the proximal and middle phalanges can be palpated on the dorsal surface with fingers flexed, as can the interphalangeal joints.

ARTICULAR SURFACES OF THE THUMB

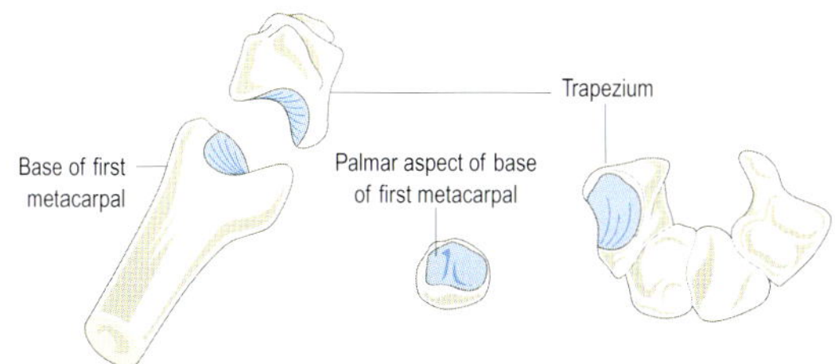

Articular surfaces of the carpometacarpal joint

Carpometacarpal joint A synovial saddle joint between the trapezium and base of the first metacarpal. The articular surface of the trapezium is concave in an anteroposterior plane and convex in a transverse plane: the base of the metacarpal is reciprocally curved.

The joint provides a stable base from which the thumb can work effectively and efficiently.

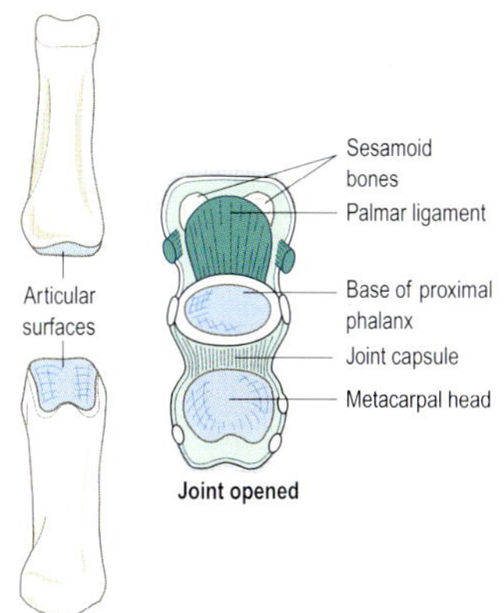

Metacarpophalangeal joint

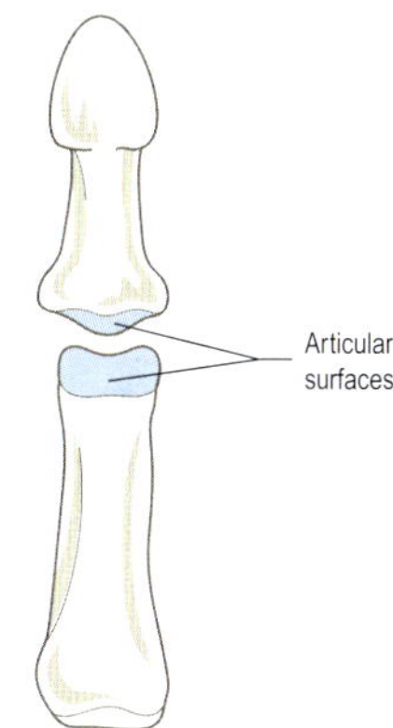

Interphalangeal joint

Metacarpophalangeal joint A synovial condyloid joint between the rounded head of the metacarpal and shallow oval concavity of the base of the proximal phalanx. The phalangeal articular surface is increased by the fibrocartilaginous palmar ligament attached to the anterior surface of the base of the phalanx.

Interphalangeal joint A synovial hinge joint between the pulley-shaped head of the proximal phalanx and the base of the distal phalanx. The distal phalangeal articular surface is increased by the fibrocartilaginous palmar ligament attached to the anterior surface of the base of the distal phalanx.

CAPSULE AND LIGAMENTS OF JOINTS OF THE THUMB

Carpometacarpal (CMC) joint A strong loose fibrous capsule completely encloses the joint attaching to the articular margins of both bones: the capsule is thickened laterally by the *radiocarpal ligament*, between the lateral surfaces of trapezium and 1st metacarpal, and *anterior* and *posterior oblique ligaments* converging from the trapezium onto the medial side of the metacarpal. The anterior oblique ligament becomes taut during extension and the posterior oblique during flexion. Synovial membrane lines all non-articular surfaces.

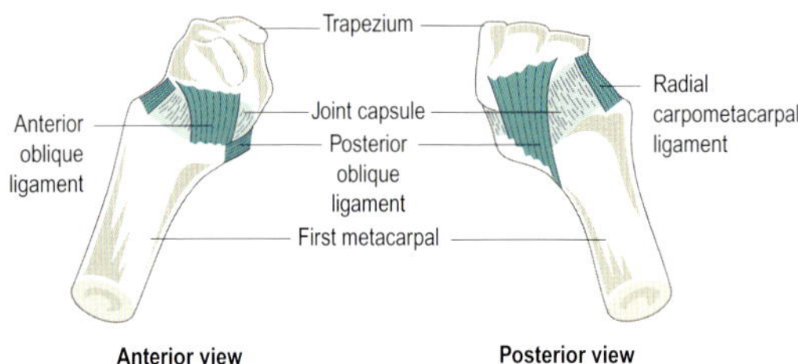

Capsule and ligaments of the CMC joint

Metacarpophalangeal (MCP) joint A loose fibrous capsule completely surrounds the joint, strengthened posteriorly by the expansion of extensor pollicis longus, at the sides by *collateral ligaments*, and replaced anteriorly by the palmar ligament. The collateral ligaments fan out from proximal to distal, also attaching to the palmar ligament.

The *palmar ligament* is attached to the anterior surface of the base of the proximal phalanx: it contains two sesamoid bones, which are attached to the phalanx and metacarpal by straight and cruciate fibres. Synovial membrane lines all non-articular surfaces.

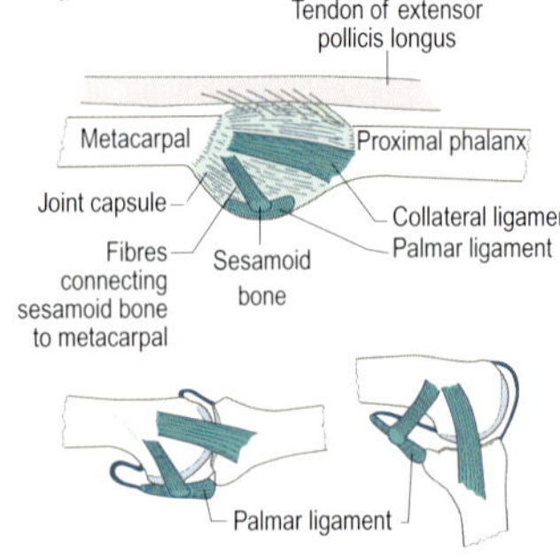

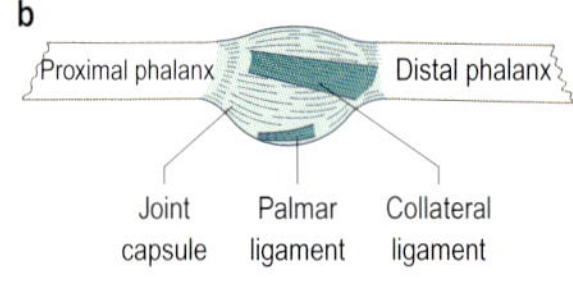

Capsule and ligaments of (a) the metacarpophalangeal and (b) the interphalangeal joints

Interphalangeal joint A fibrous capsule completely surrounds the joint, strengthened at the sides by *collateral ligaments* and replaced anteriorly by the palmar ligament. As in the MCP joint the *palmar ligament* is a fibrocartilaginous plate attached to the anterior margin of the base of the distal phalanx and loosely attached to the neck of the proximal phalanx via the joint capsule. Synovial membrane lines all non-articular surfaces.

MOVEMENTS OF THE THUMB

Because the thumb is rotated approximately 90° with respect to the plane of the hand, flexion and extension occur in a coronal plane, while abduction and adduction occur in a sagittal plane.

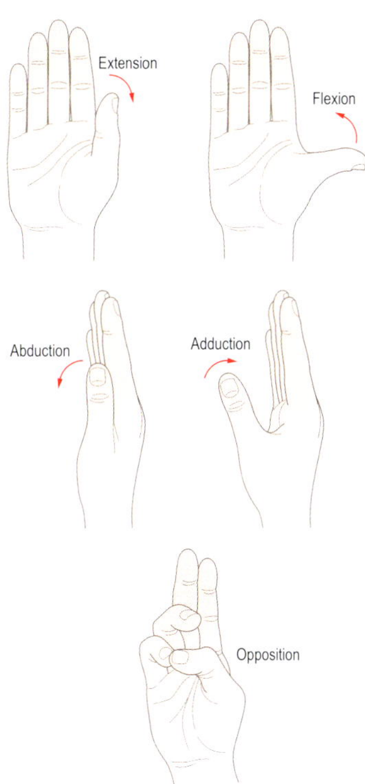

Movements of the thumb at the CMC joint

Carpometacarpal (CMC) joint *Flexion* moves the thumb medially while *extension* moves it laterally, the axis of movement being through the base of the metacarpal at the centres of curvature. The total range of movement is 40–50°.

Abduction and *adduction* occur at right angles to the palm so that in abduction the thumb is carried forwards and adduction back towards the palm. The total range of movement is 80°.

Opposition is a movement in which the distal pad of the thumb is brought against the distal pad of any of the remaining four digits. It consists of three elementary movements: simultaneous flexion and abduction, rotation and then adduction.

Accessory movements: At the CMC joint the surfaces can be moved in both anteroposterior and mediolateral directions, as well as longitudinal gapping and rotation.

MOVEMENTS OF THE THUMB

Metacarpophalangeal (MCP) joint *Flexion* and *extension* take place about a single transverse axis through the metacarpal. Flexion has a range of 45°, while extension under normal circumstances is zero. In full extension the anterior part of the metacarpal head articulates with the palmar ligament.

Abduction and *adduction* take place about an anteroposterior axis through the head of the metacarpal. Abduction has a range of 15° while adduction is negligible.

Some *axial rotation* is possible, being important during opposition of the thumb.

Interphalangeal (IP) joint *Flexion* and *extension* occur about a transverse axis through the neck of the proximal phalanx. The range of flexion exceeds 90°, while extension is limited to no more than 10°. Passive hyperextension can be marked in some individuals.

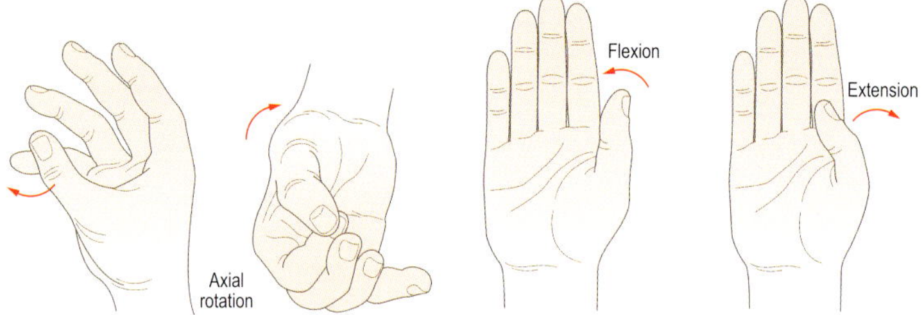

Movements at the metacarpophalangeal and interphalangeal joints

Accessory movements: Similar anteroposterior gliding movements can be elicited at the MCP and IP joints, as well as longitudinal gapping.

ARTICULAR SURFACE OF THE FINGERS

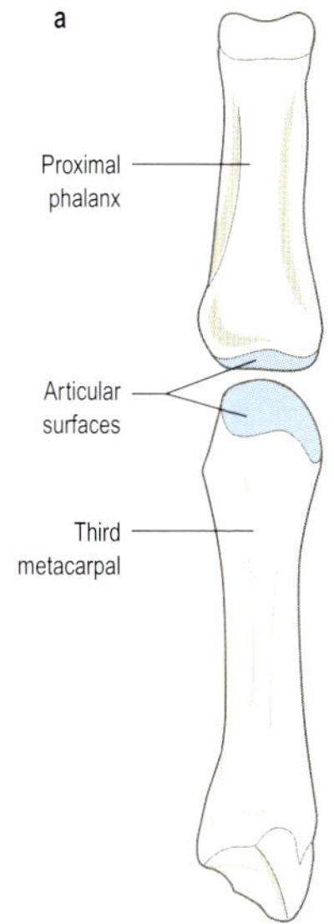

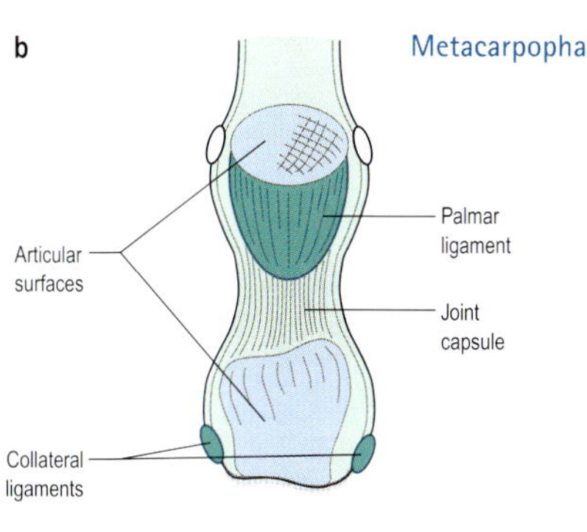

Metacarpophalangeal joint

Metacarpophalangeal (MCP) joint A synovial condyloid joint structurally and functionally similar to that of the thumb.

The metacarpal head is biconvex with unequal curvatures. The base of the proximal phalanx is biconcave but has a smaller articular surface, increased by the palmar ligament.

Interphalangeal joints Synovial hinge joints: there are two interphalangeal joints in each finger.
Proximal interphalangeal (PIP): joint between the head of the proximal phalanx and base of the middle phalanx
Distal interphalangeal (DIP): joint between the head of the middle phalanx and base of the distal phalanx.

Articulation is between the pulley-shaped head of the phalanx with two shallow facets separated by a ridge on the base of the adjacent distal phalanx. Except for joints of the index finger, in which the groove and ridge lie in an anteroposterior direction, in all others they run obliquely from posterolateral to anteromedial with the obliquity increasing from the middle to little fingers.

The mobile fibrocartilaginous palmar ligament increases the surface area of the base of the metacarpal.

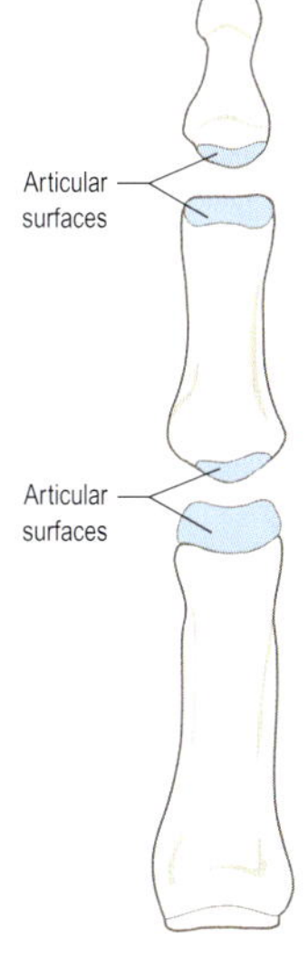

Articular surfaces of the interphalangeal joints

CAPSULE AND LIGAMENTS OF JOINTS OF THE FINGERS

Metacarpophalangeal (MCP) joint A loose fibrous capsule completely surrounds the joint, strengthened at the sides by collateral ligaments, replaced anteriorly by the palmar ligament and posteriorly by the extensor expansion, which also receives fibres from the palmar aponeurosis. *Collateral ligaments* fan out from proximal to distal also attaching to the palmar ligament. The *palmar ligament* is firmly attached to the anterior surface of the base of the proximal phalanx. Synovial membrane lines all non-articular surfaces.

Deep transverse metacarpal ligament: a series of short ligaments continuous with the palmar interosseous fascia and fibrous flexor tendons connecting the palmar ligaments of the MCP joints. Passing anterior to the ligaments are the tendons of the lumbricals and posteriorly the tendons of dorsal and palmar interossei.

Interphalangeal (IP) joints A fibrous capsule completely surrounds the joint, strengthened at the sides by *collateral ligaments* and replaced anteriorly by the palmar ligament and posteriorly by the extensor expansion. As in the MCP joint, the *palmar ligament* is a fibrocartilaginous plate attached to the anterior margin of the base of the distal phalanx and loosely attached to the neck of the proximal phalanx via the joint capsule. Synovial membrane lines all non-articular surfaces.

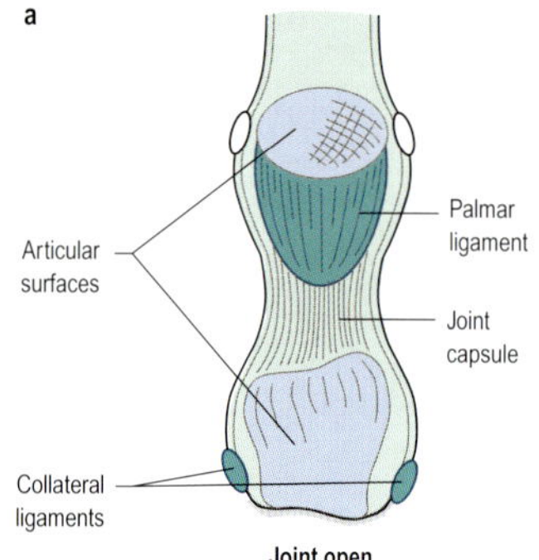

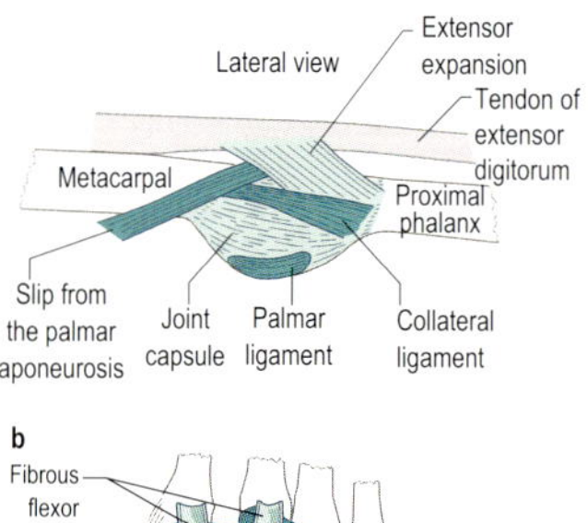

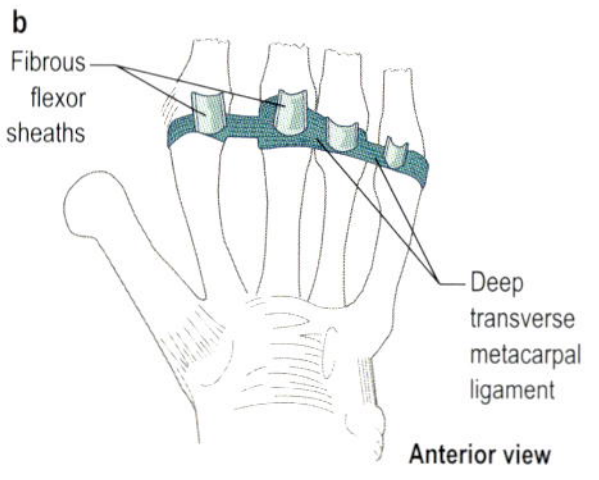

(a) Capsule and ligaments of the metacarpophalangeal joint; (b) deep transverse metacarpal ligament

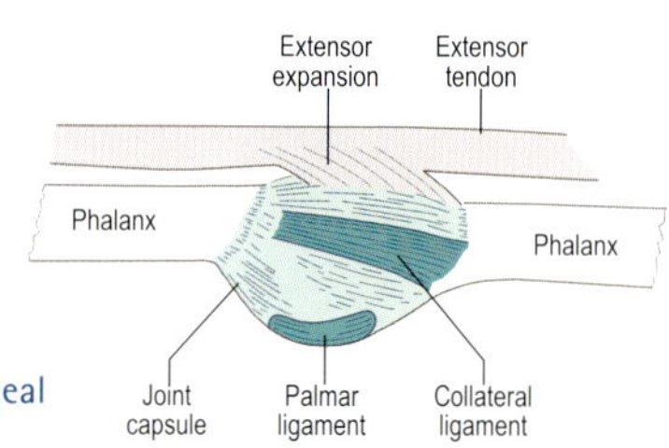

Capsule and ligaments of the interphalangeal joints

MOVEMENTS OF THE FINGERS

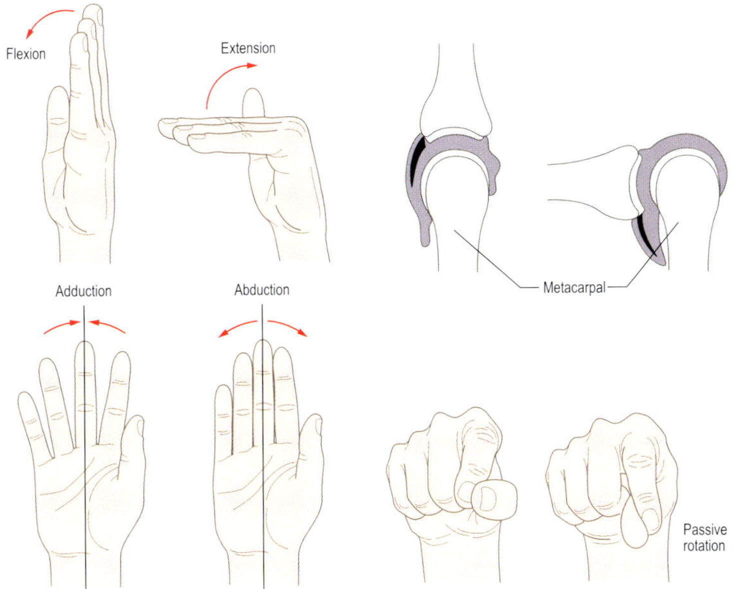

Movements at the metacarpophalangeal joints

At the MCP joints flexion and extension take place about a transverse axis and abduction and adduction about an anteroposterior axis.

Flexion/extension and abduction/adduction are both active movements; however, passive axial rotation also occurs at the MCP joint.

Metacarpophalangeal (MCP) joints In *extension* the metacarpal head articulates with the palmar ligament, while during *flexion* the ligament moves past the head. Flexion of the index finger is approximately 90°, increasing towards the little finger: extension is variable but may reach 50° in some individuals, with passive extension reaching as much as 90°.

Abduction is movement of the index, ring and little fingers away from the middle finger, *adduction* being the reverse movement. Abduction is easier and greater with the finger extended, having a range of 30° in each direction.

MOVEMENTS OF THE FINGERS

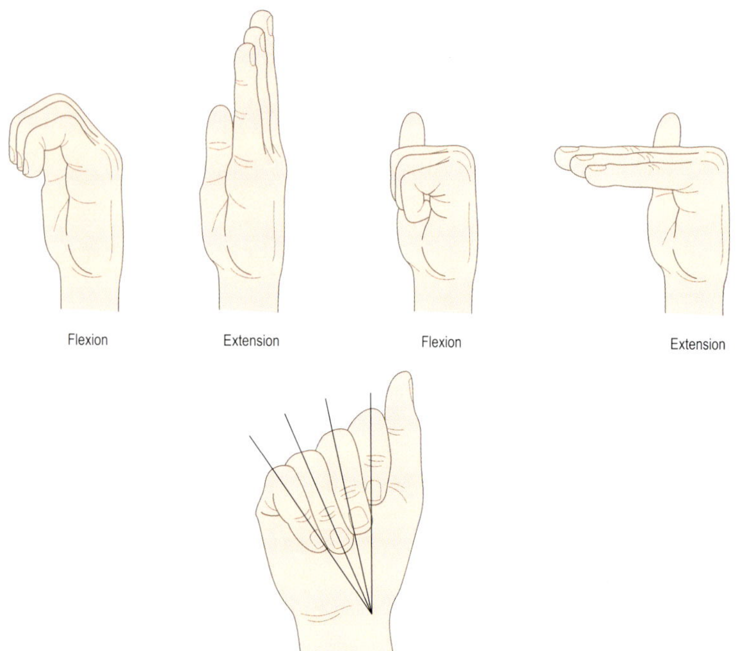

Movements at the interphalangeal joints

Interphalangeal joints *Flexion* and *extension* take place about a transverse axis which becomes increasingly oblique from the middle to little fingers.

The range of flexion at the PIP joint is 90° for the index finger, increasing to 135° for the little finger. For the DIP joints the range is 90° for the little finger, gradually decreasing to the index finger.

Accessory movements: Anteroposterior gliding and rotation are possible at both the MCP and IP joints of all four fingers.

MUSCLES

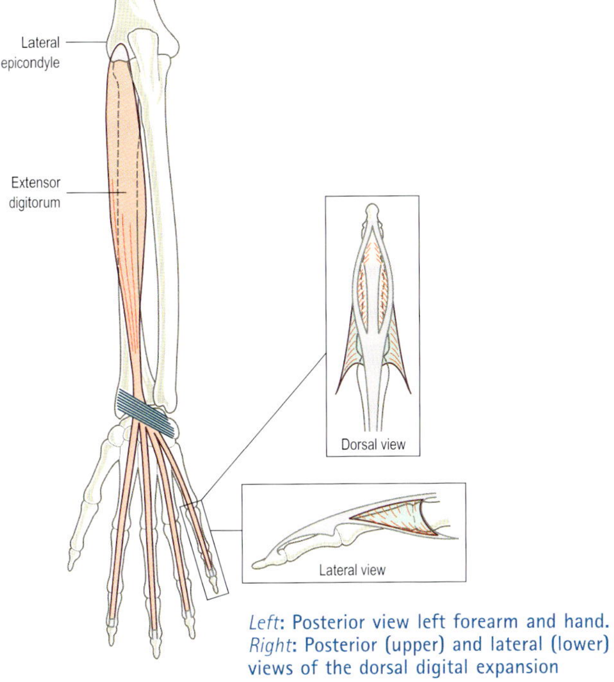

Left: Posterior view left forearm and hand. *Right*: Posterior (upper) and lateral (lower) views of the dorsal digital expansion

Extensor digitorum – extends in turn: distal interphalangeal (DIP), proximal interphalangeal (PIP), metacarpophalangeal (MCP) and wrist joints.
Origin: lateral epicondyle humerus via common extensor tendon.
Insertion: via dorsal digital expansion to base of posterior surface distal and middle phalanges of fingers.
Nerve supply: posterior interosseous branch of radial nerve C7, 8.

The dorsal digital expansion (or extensor hood) is a moveable triangular fascial structure on the posterior surface of each finger. Its central portion is formed by the tendon of extensor digitorum which splits into three, as shown above. The 'wings' at the base of the expansion wrap around the finger on to the palmar surface where the lumbricals attach. The tendons of extensors indicis and digiti minimi also attach to the dorsal digital expansion.

Rupture of the attachment to the distal phalanx results in 'mallet finger'. Rupture of the attachment to the middle phalanx with the distal still intact produces a 'boutonnière' deformity, where the PIP joint protrudes posteriorly between the tendon(s).

MUSCLES

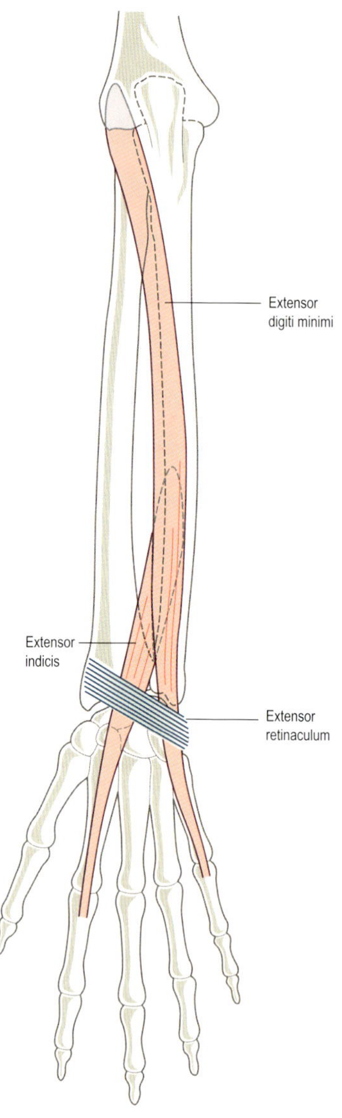

Posterior view left forearm and hand

Extensor indicis – extends index finger.

Origin: lower part posterior surface of ulna and adjacent interosseous membrane.

Insertion: joins dorsal digital expansion of the index finger.

Nerve supply: posterior interosseous branch of radial nerve C7, 8.

Extensor digiti minimi – extends little finger.

Origin: lateral epicondyle of humerus via common extensor tendon.

Insertion: joins dorsal digital expansion of little finger.

Nerve supply: posterior interosseous branch of radial nerve C7, 8.

The four tendons of extensor digitorum usually act together; however, extensors indicis and digiti minimi allow these fingers to move independently of the others.

All six tendons pass deep to the extensor retinaculum where they are enclosed in synovial sheaths. These sheaths do not extend into the fingers.

MUSCLES

Flexor digitorum superficialis – flexes in turn: PIP, MCP joints of the four fingers and the wrist joint; is superficial to flexor digitorum profundus.

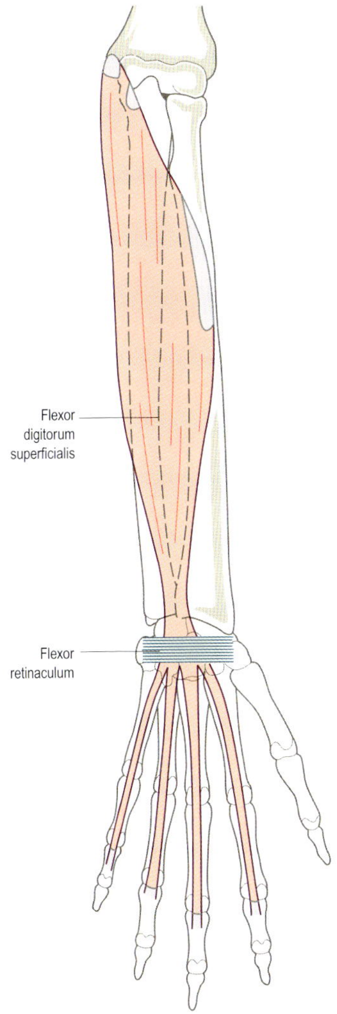

Anterior view left forearm and hand

Origin: medial epicondyle humerus via common flexor tendon, ulnar collateral ligament, medial part coronoid process, upper two-thirds anterior border of radius.

Insertion: palmar surface base of middle phalanx of all fingers.

Nerve supply: median nerve C7, 8, T1.

Below the flexor retinaculum the tendons of superficialis lie above profundus as two pairs lying on top of one another. At the level of the MCP joints superficialis tendons split and pass around profundus tendons before rejoining and inserting into the middle phalanx. Profundus tendons, having passed through superficialis, continue to their insertion on the distal phalanx (see diagram below).

In the hand superficialis and profundus tendons share a synovial sheath and are held in place by fibro-osseous canals.

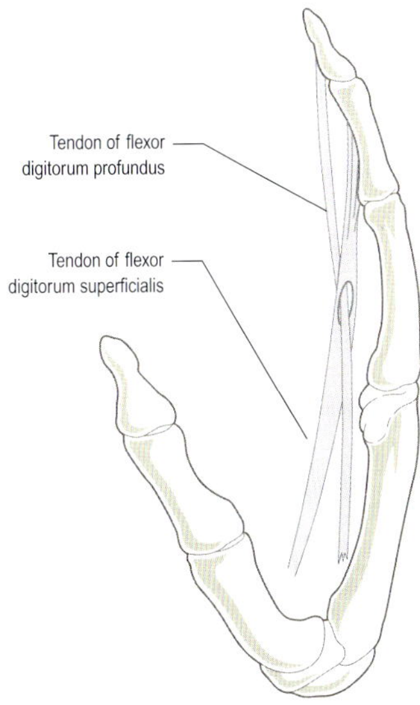

Relationship of long flexor tendons

MUSCLES

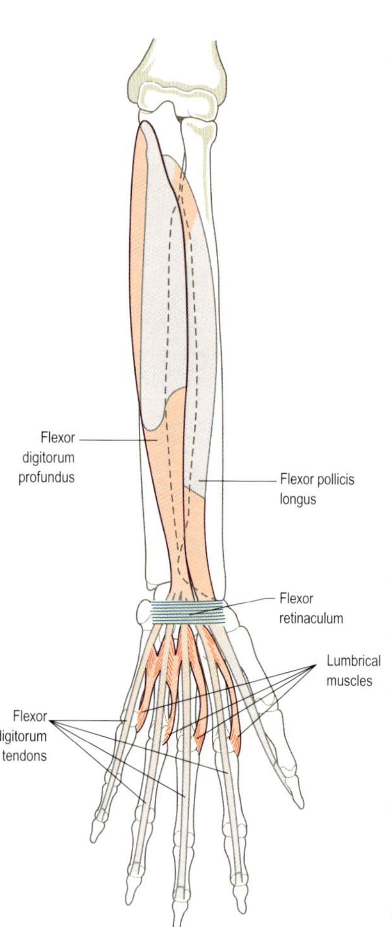

Anterior view left forearm and hand

Flexor digitorum profundus – flexes in turn: distal, proximal interphalangeal joints (DIP, PIP), metacarpophalangeal (MCP) and wrist joints.
Origin: medial side coronoid process and upper three-quarters anterior and medial surfaces ulna, middle third anterior surface interosseous membrane, posterior border ulna via aponeurosis.
Insertion: palmar surface base distal phalanx of all fingers.
Nerve supply: index and middle fingers – anterior interosseous branch median nerve C7, 8, T1; little and ring fingers – ulnar nerve C8, T1.

The four tendons of profundus pass below the flexor retinaculum lying side by side and deep to the superficialis tendons with which they share a synovial sheath.

Lumbricals – flex MCP joint and extend DIP and PIP joints of all fingers.
Origin: lateral side of flexor digitorum profundus tendons.
Insertion: lateral edge dorsal digital expansion of same finger.
Nerve supply: lateral two – median nerve T1; medial two – ulnar nerve T1.

Flexor pollicis longus – the only flexor of the distal phalanx of thumb.
Origin: anterior surface of radius between radial tuberosity and pronator quadratus, and adjacent interosseous membrane.
Insertion: palmar surface base of distal phalanx of thumb.
Nerve supply: anterior interosseous branch, median nerve C8, T1.

The tendon passes below the flexor retinaculum in its own synovial sheath.

MUSCLES

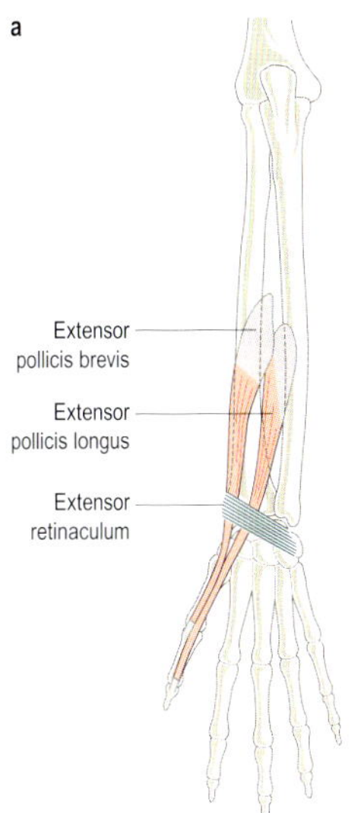

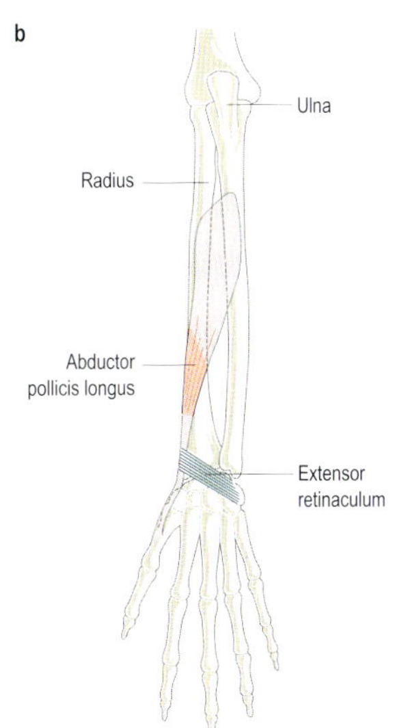

Posterior views left forearm and hand

Extensor pollicis longus – extends all joints of thumb and extends and abducts the wrist.

Origin: middle third posterior surface of ulna and adjacent interosseous membrane.

Insertion: dorsal surface base of distal phalanx of thumb.

Nerve supply: posterior interosseous branch of radial nerve C7, 8.

Extensor pollicis brevis – extends carpometacarpal and metacarpophalangeal joints of thumb; also helps extend and abduct the wrist.

Origin: middle third posterior surface of radius and adjacent interosseous membrane.

Insertion: dorsal surface base of proximal phalanx of thumb.

Nerve supply: posterior interosseous branch of radial nerve C7, 8.

Abductor pollicis longus – abducts the thumb.

Origin: middle part posterior surfaces of ulna and radius and adjacent interosseous membrane.

Insertion: radial side, base of metacarpal of thumb.

Nerve supply: posterior interosseous branch of radial nerve C7, 8.

All three tendons pass deep to the extensor retinaculum enclosed in their own synovial sheaths. Extensor pollicis brevis and abductor pollicis longus run together forming the anterolateral boundary of the 'anatomical snuff-box', with extensor pollicis longus forming the posteromedial border.

> Extensor pollicis longus tendon is prone to rupture as it crosses the radius around the dorsal tubercle.
>
> Inflammation of the sheaths surrounding extensor pollicis brevis and extensor pollicis longus (DeQuervain's tenosynovitis) may result in difficulty extending and abducting the thumb.

MUSCLES

Flexor pollicis brevis – flexes metacarpophalangeal and carpometacarpal joints of thumb.

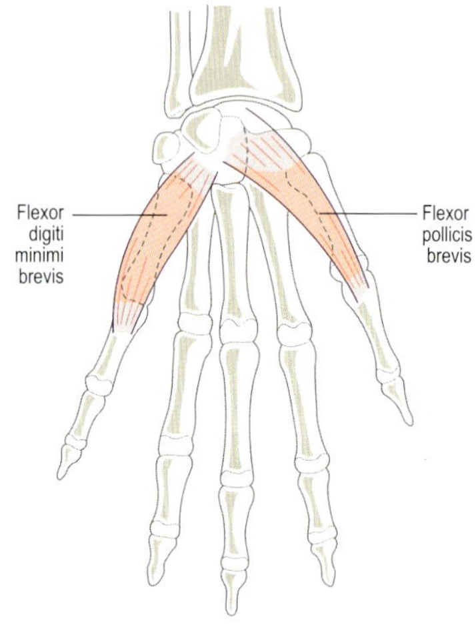

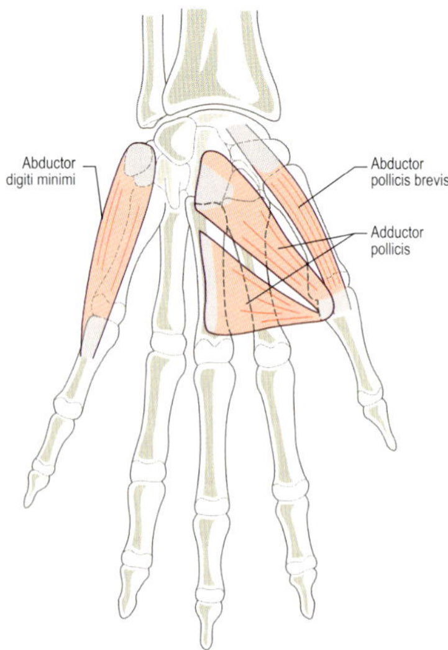

Anterior views left hand

Origin: tubercle of trapezium; capitate and trapezoid; flexor retinaculum.
Insertion: radial side base of proximal phalanx of thumb.
Nerve supply: median nerve T1 (with occasional additional ulnar nerve supply).

Flexor digiti minimi brevis – flexes metacarpophalangeal joint of little finger.
Origin: hook of hamate and flexor retinaculum.
Insertion: base proximal phalanx of little finger.
Nerve supply: ulnar nerve T1.

Abductor pollicis brevis – abducts thumb.
Origin: tubercles of scaphoid and trapezium and flexor retinaculum.
Insertion: radial side base of proximal phalanx of thumb.
Nerve supply: median nerve T1.

Abductor digiti minimi – abducts little finger.
Origin: pisiform and adjacent ligaments.
Insertion: ulnar side proximal phalanx of little finger and dorsal digital expansion.
Nerve supply: ulnar nerve T1.

Abductor digiti minimi has a role in stabilizing ulnar side of hand on supporting surface when performing precision tasks such as writing, fine painting, etc.

Adductor pollicis – adducts thumb.
Origin: oblique head – from bases 2nd, 3rd, 4th metacarpals, trapezoid and capitate; transverse head – palmar surface, shaft of 3rd metacarpal.
Insertion: medial side base of proximal phalanx of thumb by a tendon containing a sesamoid bone.
Nerve supply: ulnar nerve C8, T1.

MUSCLES

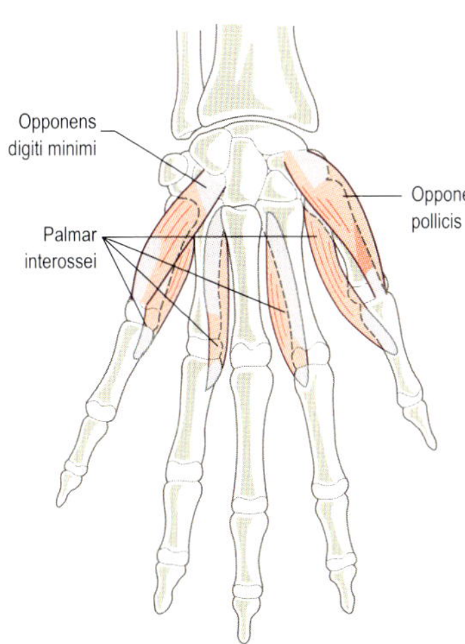

Anterior view left hand

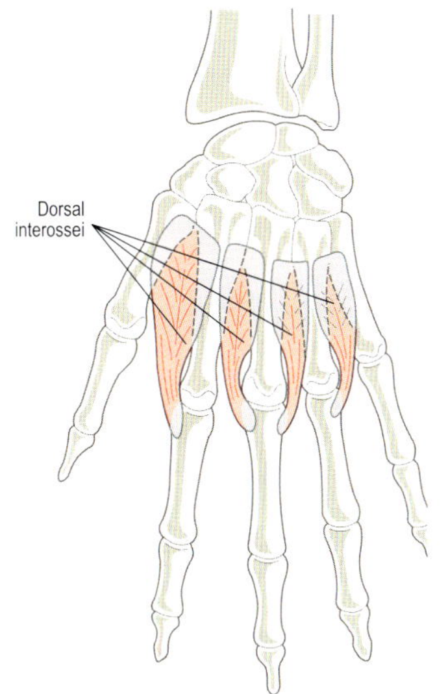

Posterior view left hand

Opponens pollicis – opposes thumb.
Origin: tubercle of trapezium and flexor retinaculum.
Insertion: whole length of palmar surface 1st metacarpal.
Nerve supply: median nerve T1.

Opponens digiti minimi – opposes little finger.
Origin: hook of hamate and flexor retinaculum.
Insertion: palmar surface 5th metacarpal.
Nerve supply: ulnar nerve T1.

Palmar interossei – adduct all digits except middle finger.
Origin and insertion: each muscle arises from the shaft of a metacarpal and inserts into the dorsal digital expansion and base of proximal phalanx of the same digit.
Nerve supply: ulnar nerve T1.

Dorsal interossei – abduct the index, middle and ring fingers.
Origin and insertion: as bipennate muscles they attach to sides of adjacent metacarpals and insert into the dorsal digital expansion and proximal phalanx of the appropriate finger.
Nerve supply: ulnar nerve T1.

Abduction of the fingers and thumb is the movement away from a line drawn down the centre of the middle finger. Movement of the middle finger in either direction is abduction, explaining why two dorsal but no palmar interossei attach to this finger. Adduction is movement towards the middle finger.

Opposition of little finger and thumb is a combined movement where they swing into the palm. Opposition of thumb allows contact with any other digit.

SYNOVIAL SHEATHS

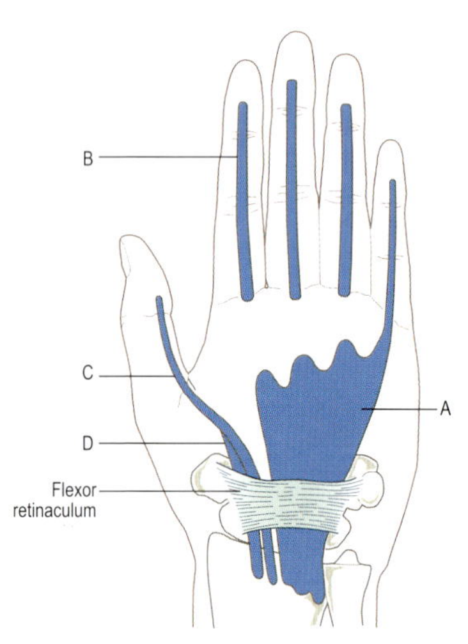

Passing deep to the flexor retinaculum the tendons of flexors digitorum superficialis and profundus lie in the same synovial sheath, with the tendons of flexor pollicis longus and flexor carpi radialis in their own sheaths laterally: all synovial sheaths extend for a short distance proximal to the flexor retinaculum. The synovial sheath surrounding the flexor tendons to the little finger continues into the fibrous flexor sheath of that finger, while the synovial sheaths lining the fibrous flexor sheaths of the index, middle and ring fingers do not communicate with the synovial sheath in the palm.

Synovial sheaths of the anterior wrist and palm: A – flexors digitorum superficialis and profundus; B – flexor tendon in the fingers; C – flexor pollicis longus; D – flexor carpi radialis

As the extensor tendons pass into the hand they are enclosed in synovial sheaths, which start proximal to the extensor retinaculum. From lateral to medial the tendons of abductor pollicis longus and extensor brevis share the same sheath, as do the tendons of extensors carpi radialis longus and brevis. Extensor pollicis longus has its own sheath, the tendons of extensor digitorum and extensor indicis share a sheath, while the tendons of extensors digiti minimi and carpi ulnaris have separate sheaths.

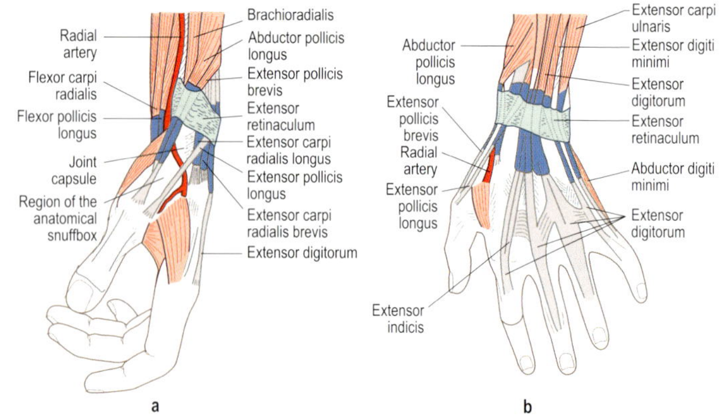

(a) Lateral and (b) posterior relations of tendons, nerves and vessels of the hand

CONNECTIVE TISSUES

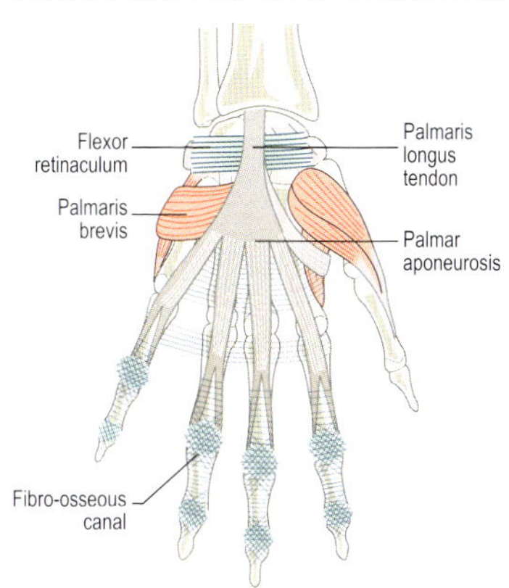

Palmar view left hand

Palmaris brevis – closely associated with fascia in the hand and assists wrinkling of the ulnar side of palm to enhance grip.
Origin: medial border palmar aponeurosis; flexor retinaculum.
Insertion: skin on medial side of hand.
Nerve supply: ulnar nerve T1.

Palmar aponeurosis A triangular sheet of dense fibrous tissue covering all the long muscle tendons in the palm. The apex is joined to the flexor retinaculum at the wrist and may receive the insertion of palmaris longus. From the base at the webs of the fingers, four slips pass distally to become continuous with the fibro-osseous canals of the flexor tendons. It also has connections to the capsule and ligaments of the metacarpophalangeal joints.

The flexor tendons are subject to inflammation, *tendinitis*, often the result of repetitive (cumulative) minor trauma. The synovial sheaths surrounding tendons at the wrist may be affected in a similar way resulting in *tenosynovitis*.

In *Dupuytren's contracture* the medial portion of the palmar aponeurosis and fibrous sheaths of ring and little fingers become thickened and shorten, pulling these fingers down into the palm. This can cause problems with hand function.

Trigger finger results when the flexor digitorum profundus tendon is subjected to friction where it enters the tendon sheath and produces localized swelling of the tendon. Movement of the tendon is restricted at this point and may then 'click' and free itself as it enters the tendon sheath, producing sudden movement of the finger. The same condition may also affect the long flexor tendon in the thumb.

RELATIONS

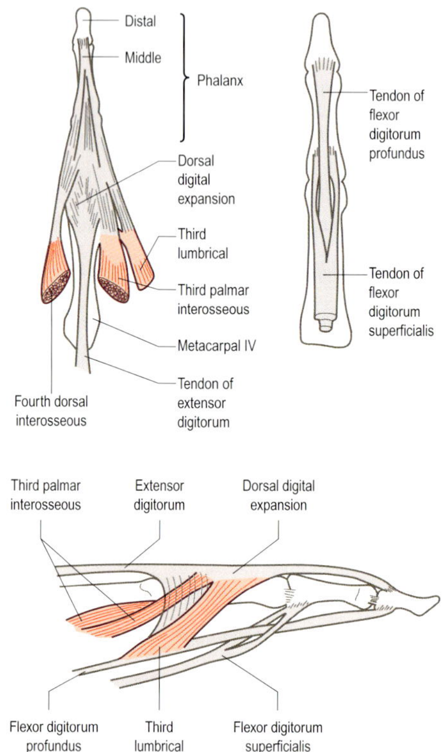

Relations of metacarpophalangeal joint

Metacarpophalangeal joint The extensor expansion covers the posterior aspect passing around the sides of the metacarpal head to blend with the deep transverse metacarpal ligament. The lumbricals pass laterally anterior and the interossei posterior to the deep transverse metacarpal ligament: which interossei pass medially and laterally depends on the finger under consideration.

Immediately anterior is the tendon of flexor digitorum profundus, with that of flexor digitorum superficialis anterior to that. Flexor digiti minimi lies anterolateral to the joint of the little finger.

Digital branches from the palmar and dorsal metacarpal arteries, together with digital branches from the median, ulnar and radial nerves (depending on finger) pass either side of the joint.

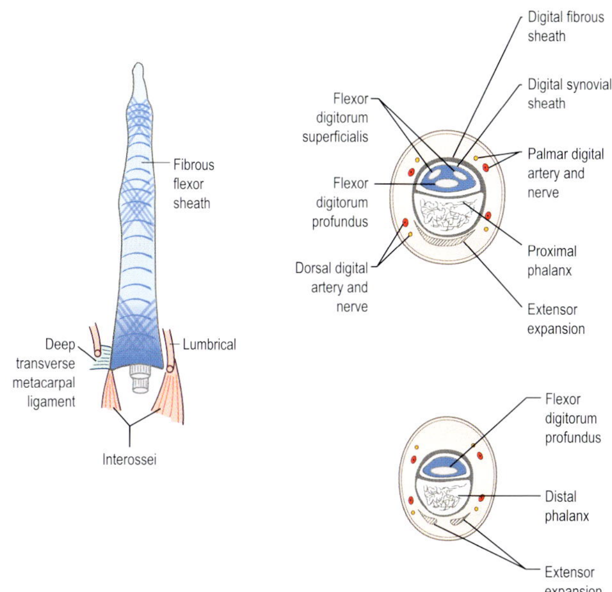

Relations of interphalangeal joints

Interphalangeal joints Posterior to the PIP is the central slip of the extensor expansion; at the DIP the two collateral slips have joined to form a single tendon.

Anterior to the PIP within the fibrous flexor sheaths are the tendons of flexors digitorum superficialis and profundus: the tendon of superficialis splits at the level of the PIP. Only the tendon of profundus lies anterior to the DIP. The fibrous flexor sheaths are relatively thin and loose over the IP joints: immediately beyond the DIP they attach to the palmar surface of the distal phalanx.

Digital branches from the palmar and dorsal metacarpal arteries, together with digital branches from the median, ulnar and radial nerves (depending on which finger), pass either side of the joints.

GRIPS

Grips can be classified as being either precision or power depending on the size, shape and weight of the object held, and the use to which it is put.

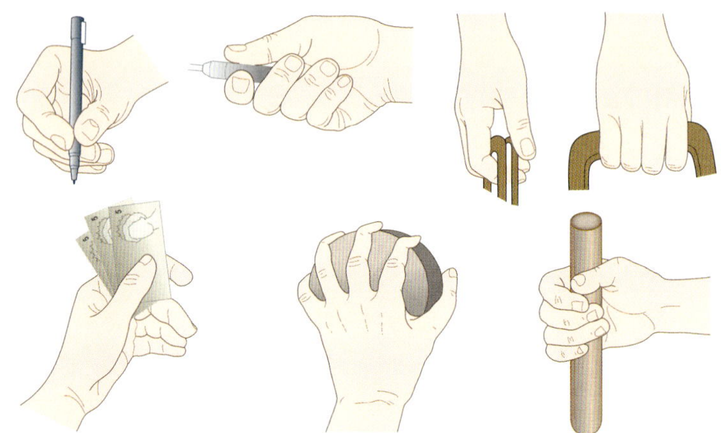

A range of grips demonstrating functions of the hand

In *precision grips* the object is held between the pads of the digits. This involves rotation at the CMC and MCP joints of the thumb and the MCP joint of the fingers, using the intrinsic muscles of the hand and flexors digitorum superficialis, profundus and pollicis longus.

Terminal opposition (pincer grip): the most precise of all precision grips in which the tips of the pads or edges of the nails are used.

Subterminal opposition (pinch grip): commonest form of precision grip using the palmar surfaces of the thumb and finger.

Subtermino-lateral opposition (key grip): less fine but strong grip where the pad of the thumb presses against the side of any of the phalanges of the finger: when the distal phalanx of the index finger is lost it can replace terminal and subterminal opposition.

Power grips are used when considerable power is required and make use of the long flexors and extensors partly fixing the wrist and partly gripping the object.

Palmar grip: the most powerful grip in which the thumb acts as a buttress and the fingers wrap around the object: the volume of the object determines the strength of the grip, which is greatest when the tips of the index finger and thumb can still touch.

Hook grip: one in which the object is held firmly between the palm and flexed fingers.

Span grip: uses the abducted fingers and thumb in addition to the palm in order to grip circular objects, e.g. jar lids.

SUMMARY – MUSCLES AND MOVEMENTS

PECTORAL GIRDLE

Retraction
Rhomboid major and minor
Trapezius

Protraction
Serratus anterior
Pectoralis minor

Elevation
Trapezius (upper fibres)
Levator scapulae

Depression
Pectoralis minor
Trapezius (lower fibres)

Lateral rotation
Trapezius
Serratus anterior

Medial rotation
Rhomboid major and minor
Levator scapulae
Pectoralis minor

SHOULDER

Abduction
Supraspinatus
Deltoid

Adduction
Coracobrachialis
Pectoralis major
Latissimus dorsi
Teres major

Flexion
Pectoralis major
Deltoid (anterior fibres)
Coracobrachialis
Biceps brachii (long head)

Extension
Latissimus dorsi
Teres major
Pectoralis major (to midline)
Deltoid (posterior fibres)
Triceps (long head)

Medial rotation
Subscapularis
Teres major
Latissimus dorsi
Pectoralis major
Deltoid (anterior fibres)

Lateral rotation
Teres minor
Infraspinatus
Deltoid (posterior fibres)

ELBOW

Flexion
Biceps brachii
Brachialis
Brachioradialis
Pronator teres

Extension
Triceps
Anconeus

FOREARM

Supination
Supinator
Biceps brachii
Brachioradialis

Pronation
Pronators teres and quadratus
Brachioradialis

WRIST

Flexion
Flexors carpi ulnaris and radialis
Palmaris longus
Flexors of fingers and thumb*

Extension
Extensors carpi radialis longus, brevis and ulnaris
Extensors of fingers and thumb*

Abduction (radial deviation)
Flexor carpi radialis
Extensors carpi radialis longus and brevis

Adduction (ulnar deviation)
Flexor carpi ulnaris
Extensor carpi ulnaris

FINGERS AND THUMB

Flexion
Flexor digitorum profundus (DIP, PIP, MCP)
Flexor digitorum superficialis (PIP, MCP)
Lumbricals and interossei (MCP)
Flexor pollicis longus (IP, MCP)
Flexor pollicis brevis (MCP)

Extension
Extensors digitorum, indicis and digiti minimi (MCP, PIP, DIP)
Extensor pollicis longus (IP, MCP)
Extensor pollicis brevis (MCP)

Abduction
Dorsal interossei
Abductor digiti minimi
Abductors pollicis longus and brevis

Adduction
Palmar interossei
Adductor pollicis

Opposition
Opponens pollicis and digiti minimi

*In continued action after moving fingers

DIP – distal interphalangeal; IP – interphalangeal; MCP – metacarpophalangeal; PIP – proximal interphalangeal

NERVE SUPPLY

BRACHIAL PLEXUS

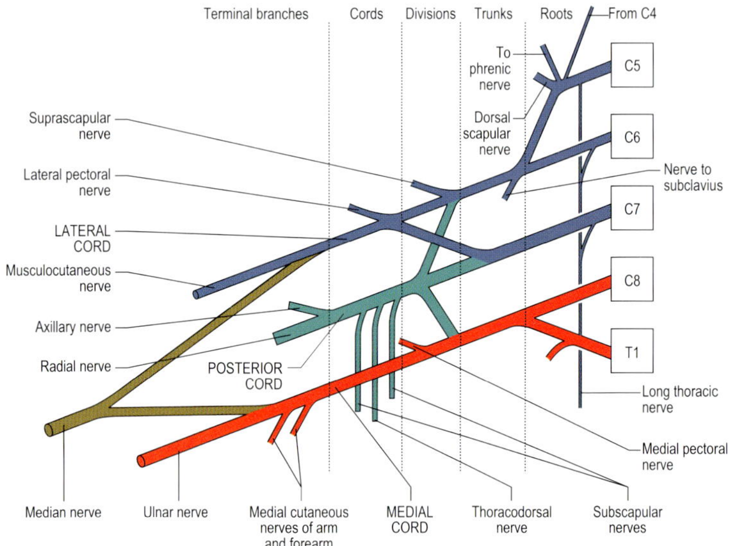

Damage to the brachial plexus can result from violent traction injuries around the shoulder joint and/or lateral cervical spine. Upper cords of the plexus may be damaged, resulting in *Erb's palsy.* Forced flexion of the neck and depression of the shoulder may occur in motorcycle injuries or, more rarely, at birth due to a difficult delivery. This type of palsy mainly affects the upper arm, elbow and wrist extensors. Damage to the lower cords results in *Klumpke's palsy*, mainly affecting the hand, and can follow violent abduction of the arm (as in grabbing for support during fall from a height) or anterior dislocation of the shoulder. It is possible for all roots to be affected, resulting in paralysis and loss of sensation throughout the upper limb, giving a 'flail arm'.

AXILLARY AND MUSCULOCUTANEOUS NERVES

Axillary nerve

Cutaneous supply:

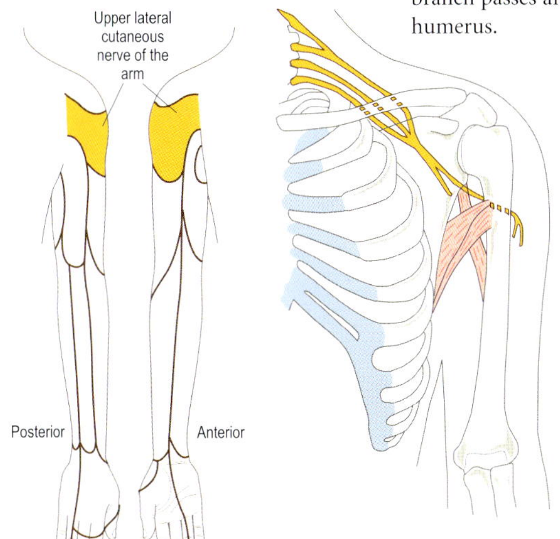

Origin: posterior cord brachial plexus C5, 6.

Course: inferior to shoulder joint, through the quadrilateral space; anterior branch passes around surgical neck of humerus.

Muscle supply:

- Deltoid
- Teres minor

Site of injury

Below shoulder joint – dislocation of shoulder, fracture surgical neck of humerus or upward pressure into axilla.

Musculocutaneous nerve

Cutaneous supply:

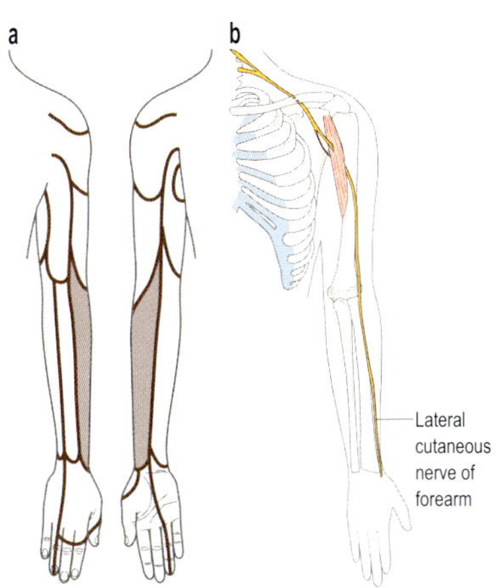

Origin: lateral cord, brachial plexus C5, 6, 7.

Course: pierces coracobrachialis, continues between biceps and brachialis to lateral side of arm, below elbow becomes lateral cutaneous nerve of forearm.

Muscle supply:

- Coracobrachialis
- Biceps brachii
- Medial two-thirds brachialis

ULNAR NERVE

Origin: medial cord brachial plexus (C7) C8, T1.
Course: pierces intermuscular septum to run behind medial epicondyle of humerus in ulnar groove; enters anterior forearm between two heads of flexor carpi ulnaris and runs deep to flexor carpi ulnaris; enters hand superficial to flexor retinaculum.
Muscle supply:
Forearm:

- Flexor carpi ulnaris
- Flexor digitorum profundus (little and ring fingers)

Hand:

- Palmaris brevis
- Abductor digiti minimi
- Flexor digiti minimi
- Opponens digiti minimi
- Adductor pollicis
- Medial two lumbricals
- Palmar and dorsal interossei (8)

Cutaneous distribution: palmar cutaneous, dorsal and superficial cutaneous branches.

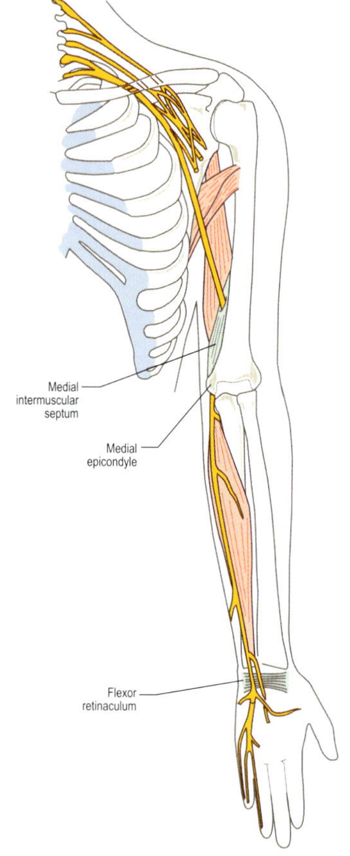

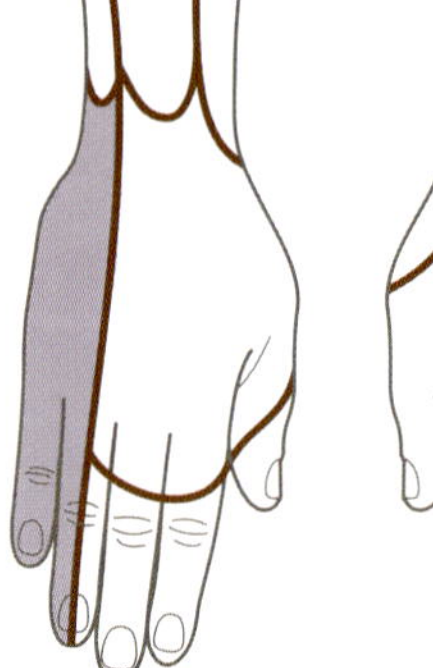

Common sites of injury

Ulnar groove – direct trauma to nerve, fracture medial epicondyle of humerus, compression due to valgus deformity.

Wrist – laceration as nerve superficial.

Typical deformity

'Claw hand' deformity in little and ring fingers due to paralysis/weakness of interossei and lumbrical muscles.

RADIAL NERVE

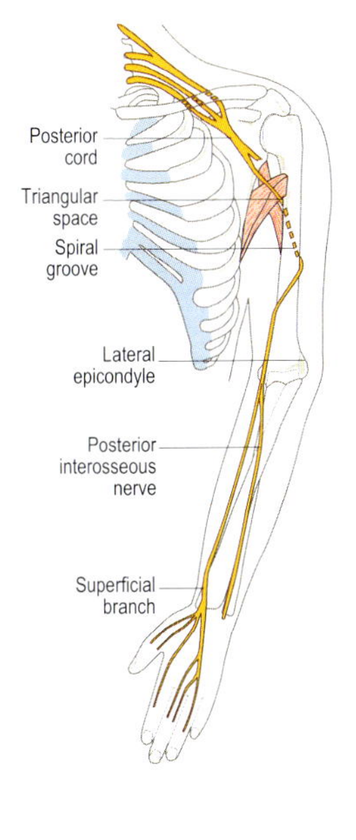

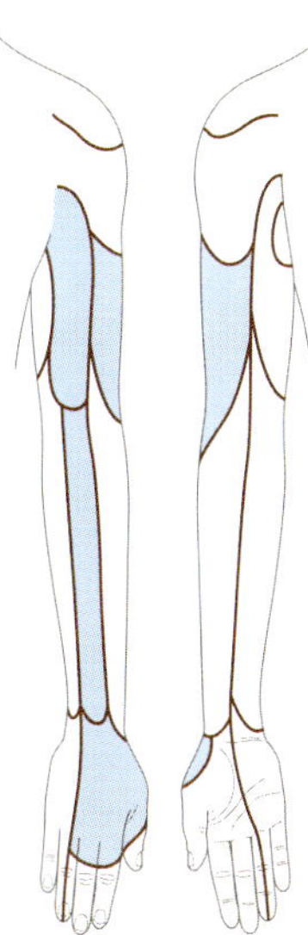

Origin: posterior cord, brachial plexus C5, 6, 7, 8 (T1).

Course: through triangular space, into spiral groove posterior aspect of humerus; pierces intermuscular septum, running between brachialis and brachioradialis; at elbow (anterior to lateral epicondyle) divides into superficial and posterior interosseous (deep) branches, the latter enters posterior forearm between heads of supinator.

Muscle supply:

Radial nerve (proper):

- Long, medial, lateral heads of triceps
- Anconeus
- One-third brachialis
- Brachioradialis
- Extensor carpi radialis longus

Posterior interosseous branch:

- Supinator
- Extensor carpi radialis brevis
- Extensor digitorum
- Extensor digiti minimi
- Extensor carpi ulnaris
- Extensors pollicis longus and brevis
- Extensor indicis
- Abductor pollicis longus

NB: Forearm is shown pronated and arm medially rotated.

Cutaneous supply:

- Posterior and lower lateral cutaneous nerves of arm
- Posterior cutaneous nerve of forearm
- Superficial branch

Common sites of injury

Axilla – compression, e.g. by axillary crutches.

Spiral groove – e.g. severance by bone ends following fracture mid-shaft of humerus, compression by callus, traumatic haematoma or even prolonged pressure of arm on back of chair.

Typical deformity

'Drop wrist'.

MEDIAN NERVE

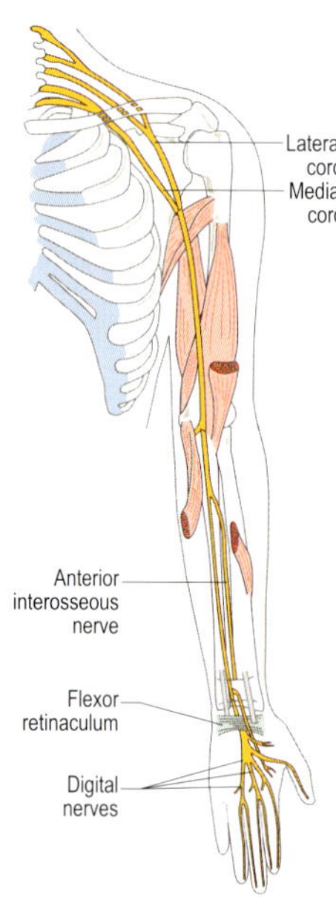

Origin:
lateral cord } brachial plexus C5, 6, 7, 8,
medial cord } T1

Course: runs deep to biceps then bicipital aponeurosis, between heads of pronator teres, deep to flexor digitorum superficialis, deep to flexor retinaculum.

Muscle supply:
- Pronator teres
- Flexor carpi radialis
- Palmaris longus
- Flexor digitorum superficialis

Via anterior interosseous nerve:
- Flexor digitorum profundus (index and middle fingers)
- Flexor pollicis longus
- Pronator quadratus

Median nerve in hand:
- Abductor pollicis brevis
- Flexor pollicis brevis
- Opponens pollicis
- Lateral two lumbricals

Common sites of injury
Elbow – supracondylar fracture.
Wrist – compression within carpal tunnel (carpal tunnel syndrome); lacerations on flexor aspect.

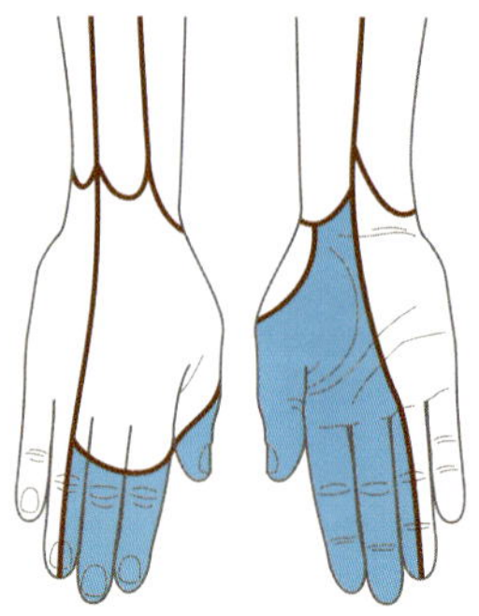

Cutaneous supply:
- Palmar cutaneous and digital nerves

CUTANEOUS NERVE SUPPLY

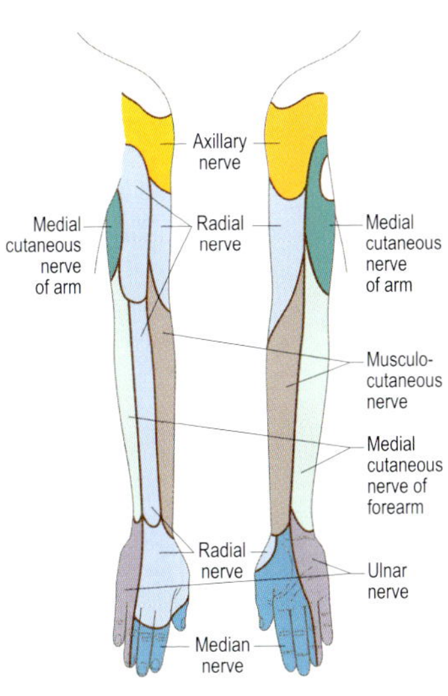

The *cutaneous distribution* of a named peripheral nerve is the area of skin supplied by this nerve. Severance of the named nerve results in a loss of sensation in the area indicated.

DERMATOMES

A *dermatome* is the area of skin innervated predominantly by one spinal nerve root. This nerve root may have contributed to a number of named peripheral nerves so severance will result in a more variable sensory loss.

BLOOD VESSELS

ARTERIES

Arteries carry blood rich in oxygen and nutrients from the heart. In the arm, the arterial system starts at the *subclavian artery* which becomes the *axillary artery* at the outer border of the 1st rib. This becomes the *brachial artery* at the lower border of teres major, dividing into *radial* and *ulnar arteries* in the cubital fossa at the level of the radial head. These terminate as the *superficial* and *deep palmar arches.*

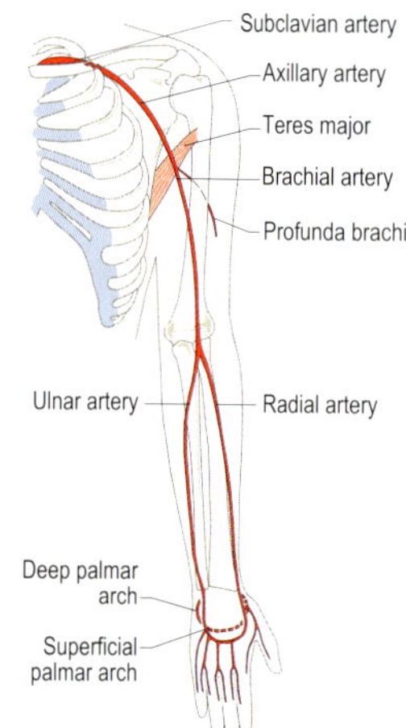

Arteries (anterior view)

Pulse points

Pulsation of an artery can be used to measure heart rate and blood pressure. The *radial pulse* is most commonly used for heart rate whilst the *brachial* is used to measure blood pressure.

VEINS

Veins carry deoxygenated blood back to the heart. In the arm they are divided into deep and superficial groups. Superficial veins start on the dorsum of the hand and continue upwards as the *basilic* and *cephalic* veins. The *anterior median vein* runs from the palm to the *median cubital vein.*

Deep veins are paired (venae comitantes) and lie alongside arteries with which they share their name. The exception is the *axillary* vein which is single and a continuation of the basilic.

Veins contain valves which only allow blood flow towards the heart. The most common point for venepuncture is the *median cubital vein.*

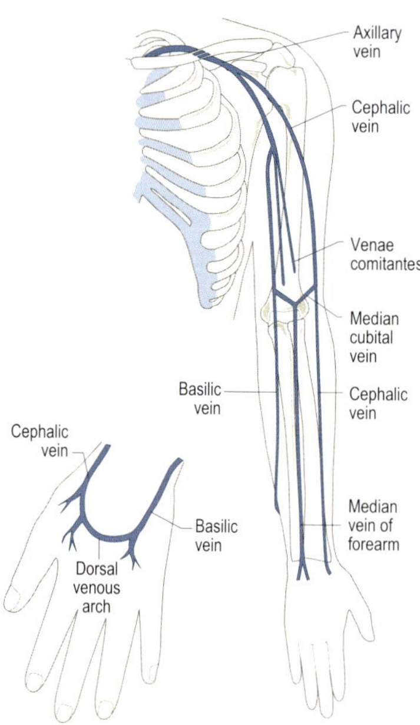

Deep and superficial veins (anterior view of arm, posterior view of hand)

PART 3

The lower limb

INTRODUCTION

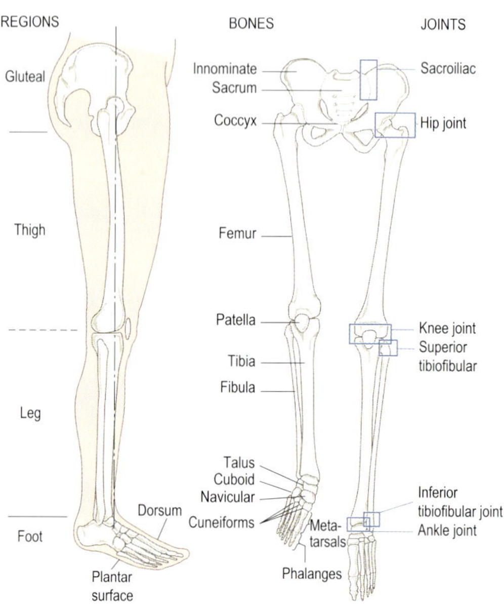

The pelvic girdle connects the lower limb to the vertebral column via the posterior articulation with the sacrum. The almost immobile sacroiliac joints provide great strength for weight transmission.

Evolutionary changes in shape and orientation of the pelvis enable the trunk to be held erect. This upright posture and bipedalism have resulted in changes to mechanical and functional requirements of the lower limb for weight-bearing and locomotion: bones are larger and more robust and joints structurally more stable than those of the upper limb.

The stable hip joint has large articular surface areas, an adaptation to increased weight-bearing, but has sacrificed mobility for stability.

The knee does not lie directly under the hip, but is closer to the midline, adding to skeletal equilibrium.

The foot has undergone the greatest evolutionary change, reflecting its role as a lever adding propulsive force to the lower limb during locomotion. Its arched arrangement converts the foot into a complex spring under tension, enabling it to check momentum at heel strike as well as transmitting the main force of thrust in forward propulsion.

The centre of gravity of the body lies close to the vertebral column, slightly behind and at the same level as the hip joint: its vertical projection passes behind the hip joint and anterior to the knee and ankle joints (as shown above).

PELVIC GIRDLE

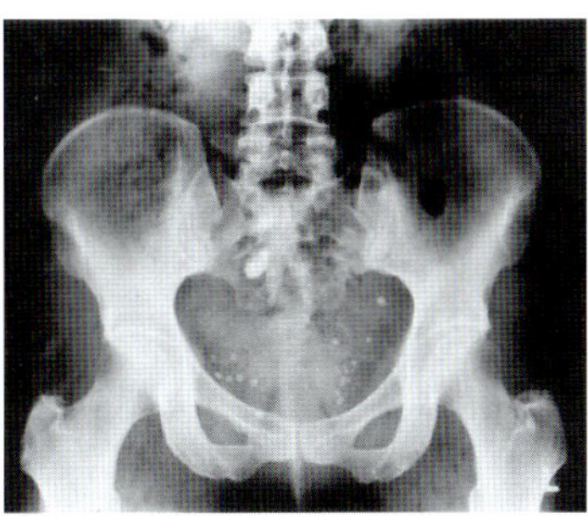

Anteroposterior radiograph of pelvis showing symphysis pubis, sacroiliac and hip joints

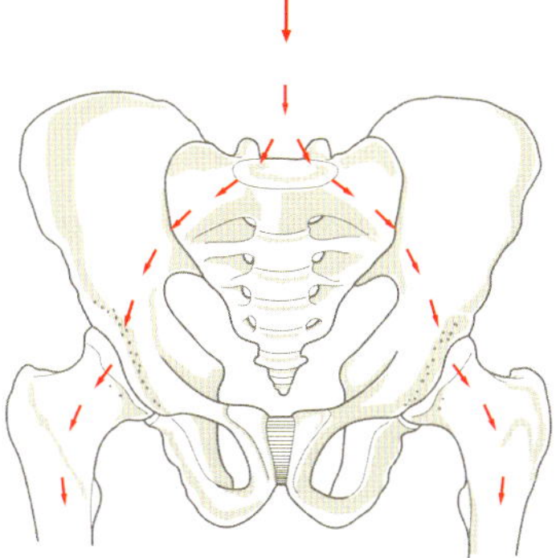

Transfer of weight from spine, through pelvis to both femora

The pelvic girdle comprises two innominate bones and the sacrum connecting the lower limb to the trunk. Each innominate articulates with the sacrum posteriorly at synovial sacroiliac joints, and with each other anteriorly at the symphysis pubis. Articulation with the lower limb is at the acetabulum and with the trunk via the sacrum, which articulates with L5 superiorly at the lumbosacral joint and coccyx inferiorly at the sacrococcygeal joint.

The lumbosacral junction is the transition between mobile and immobile parts of the vertebral column and is vulnerable to trauma and pathology.

Body weight is transferred via the pelvic girdle from the trunk to the lower limbs when standing and ischial tuberosities when sitting. Stability is achieved by loss of mobility at the sacroiliac joints and symphysis pubis.

The pelvis also supports pelvic viscera, provides extensive attachment to muscles of the trunk and lower limb and gives bony support for the birth canal in females.

BONES

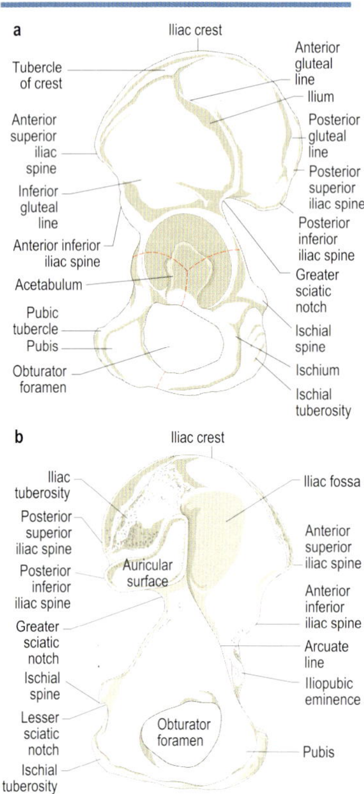

Left innominate: (a) lateral and (b) medial aspects

Innominate Irregularly shaped bone comprising the ilium, pubis and ischium fused at the acetabulum. The superior *ilium* is broad with an outer gluteal surface and inner iliac fossa; the iliac crest, which ends in anterior and posterior superior iliac spines; and a large auricular surface medially. The anterior *pubis* has a medial body and inferior and superior rami which surround the anterior part of the obturator foramen, joining the ilium and ischium respectively. The upper part of the body is the pubic crest with the pubic tubercle at its lateral end. The posterior *ischium* has a large ischial tuberosity and a blunt ischial spine, the latter separating the greater and lesser sciatic notches. The ischium completes the margin of the obturator foramen.

Palpation

The anterior superior iliac spine can be palpated anteriorly. From here the iliac crest can be traced backwards, with the iliac tubercle about 5 cm from its anterior end. At the posterior end of the crest the smaller posterior superior iliac spine can be felt.

The ischial tuberosity can be felt when sitting by placing the hand directly under the buttock.

In the lower part of the abdominal wall above the genitalia the pubic symphysis can be felt, with the pubic tubercle 1 cm above and lateral.

BONES

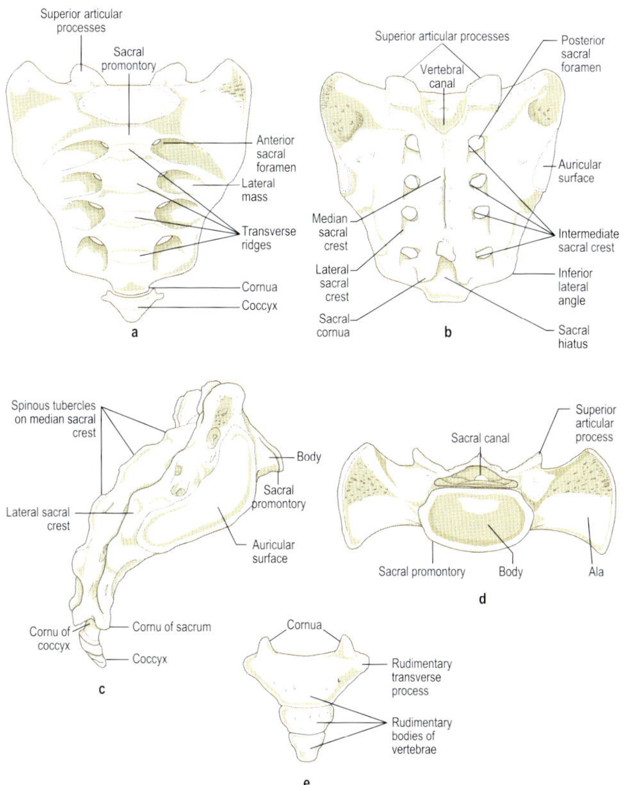

Sacrum: (a) anterior, (b) posterior, (c) lateral and (d) superior aspects; (e) coccyx

Sacrum Five fused vertebrae incorporating transverse processes (lateral masses) which extend between anterior and posterior sacral foramina. Smooth concave anterior (pelvic) surface and convex irregular dorsal surface with median, intermediate and lateral sacral crests. On the lateral aspect of the sacrum is a large auricular surface. The anterior surface of body of S1 (promontory) projects anteriorly, on each side of which are the alae. Superior articular processes of S1 articulate with inferior processes of L5: inferiorly cornua articulate with the coccyx: superiorly S1 body articulates with L5 body via L5/S1 intervertebral disc. Sacral hiatus is an inferior opening into vertebral canal.

Coccyx Four fused vertebrae: concave smooth anterior surface and convex irregular dorsal surface with row of tubercles (rudimentary articular processes).

ARTICULAR SURFACES

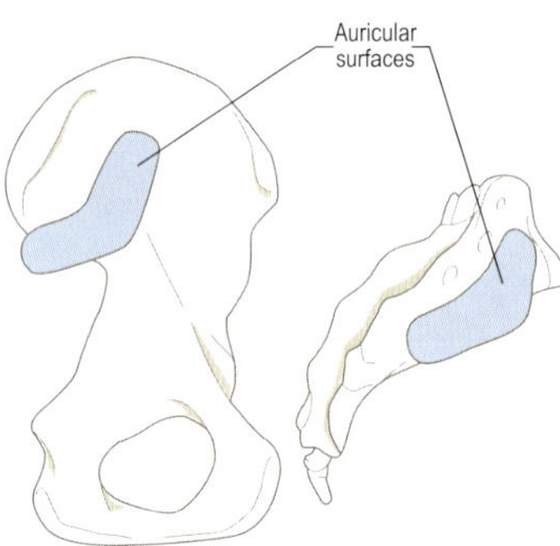

Sacroiliac joint

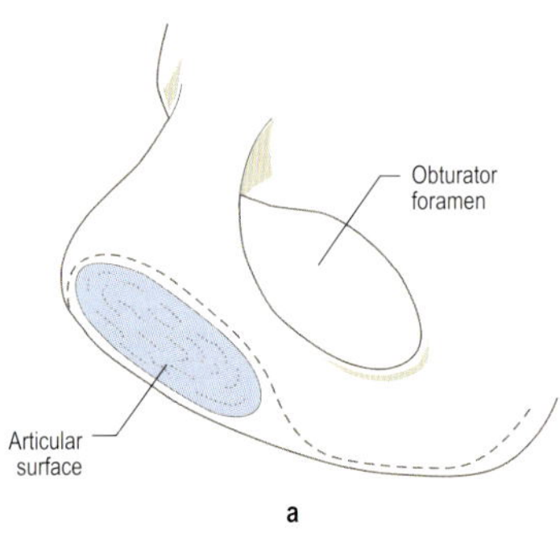

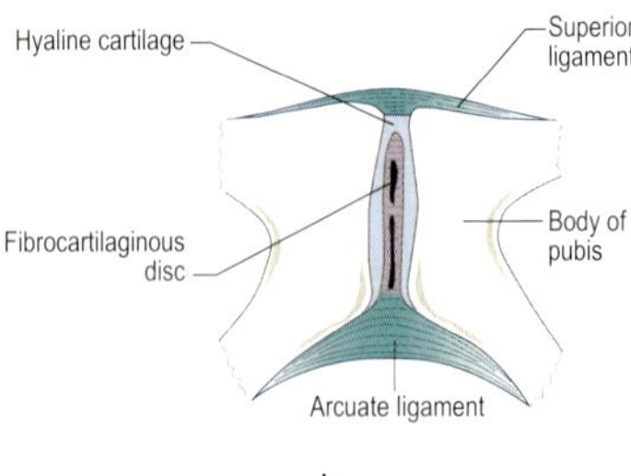

Symphysis pubis: (a) medial surface of pubic body; (b) anterior aspect

Sacroiliac joint Synovial joint between auricular surfaces of ilium and sacrum being broader above and narrower below, with marked reciprocal irregularities. The central part of the sacral surface is concave with raised crests on either side, while the ilial surface has a central crest lying between two furrows.

The shape and regularity of articular surfaces vary between individuals and between sides in the same individual. The sacral surface is covered with hyaline cartilage whereas that on the ilium is fibrocartilage. With increasing age the joint cavity becomes partially obliterated, often showing partial bony fusion in the very old.

> The *sacroiliac joint* line lies approximately 25° from vertical, passing from superolateral to inferomedial and extending 2 cm in each direction from the posterior superior iliac spine.

Symphysis pubis Secondary cartilaginous joint between the oval medial surfaces of the bodies of the pubic bones. Each surface is irregularly ridged and grooved, covered by a thin layer of hyaline cartilage and separated by a fibrocartilaginous disc, thicker in females than in males.

Sacrococcygeal joint Articulation between last sacral and first coccygeal segments via an interosseous ligament, supported by sacrococcygeal ligaments.

> **Palpation**
> The *sacrococcygeal joint* line can be felt as a horizontal groove between the apex of sacrum and the coccyx. Forward applied pressure causes rotation of coccyx against the sacrum.

LIGAMENTS ASSOCIATED WITH THE SACROILIAC JOINT

The *anterior sacroiliac ligament* lies on the pelvic surface of the joint. Broad and flat, it consists of numerous thin bands stretching from the ala and pelvic surface of the sacrum to the adjoining margin of the auricular surface of ilium. It is stronger in females.

The *posterior sacroiliac ligaments* are thicker and stronger than the anterior, filling the space between sacrum and ilium. Several distinct bands can be identified.

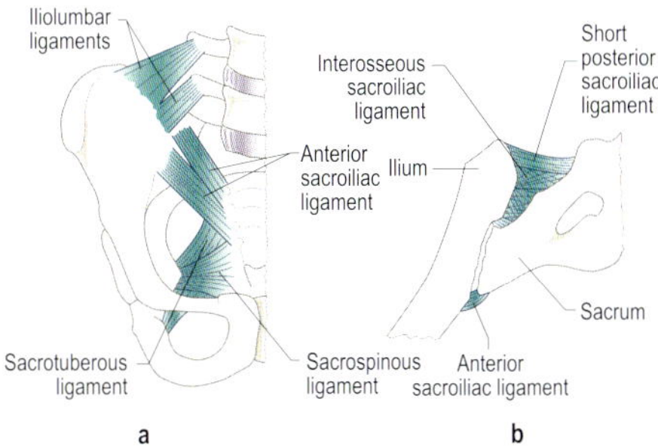

(a) Anterior sacroiliac ligament (anterior aspect); (b) sacroiliac ligaments viewed from above

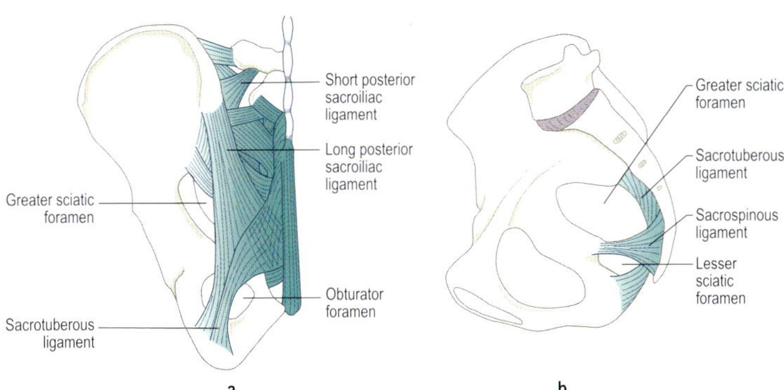

(a) Posterior sacroiliac ligaments; (b) sacrotuberous and sacrospinous ligaments

The *interosseous sacroiliac ligament* is deep, being short and thick. The horizontal fibres of the *short posterior sacroiliac ligament* lie in the upper part of the cleft and resist forward movement of the sacral promontory. Fibres of the *long posterior sacroiliac ligament* are most superficial, passing almost vertically, and resist downward movement of the sacrum with respect to the ilium.

Sacrotuberous and *sacrospinous ligaments* are accessory to the sacroiliac joint, acting to prevent forward tilting of the sacral promontory, running between the sacrum and ischial tuberosity and spine respectively.

LIGAMENTS ASSOCIATED WITH THE SYMPHYSIS PUBIS

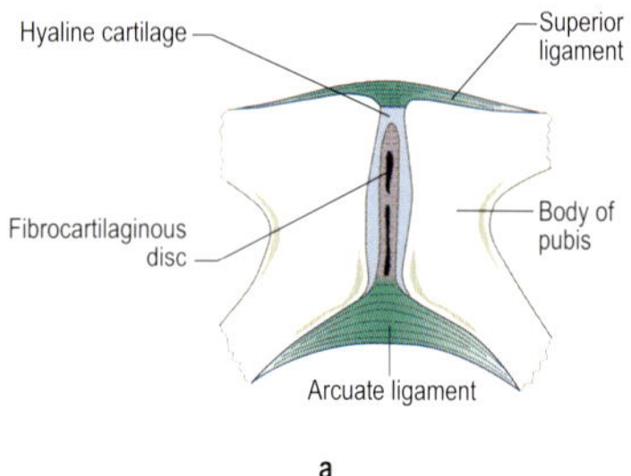

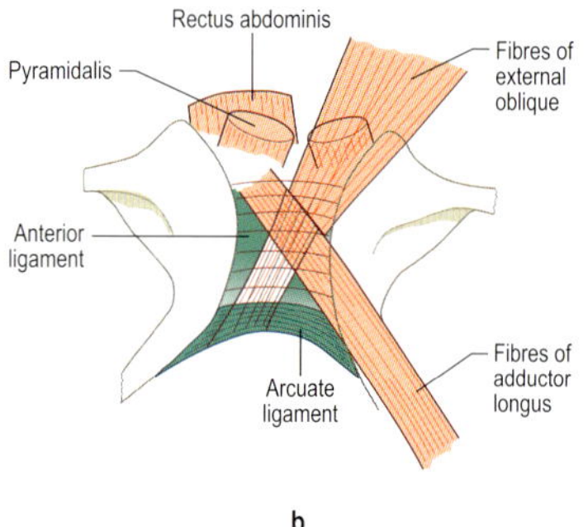

Symphysis pubis: (a) coronal section; (b) anterior aspect

The *superior pubic ligament* strengthens the anterosuperior aspect of the joint.

The *arcuate pubic ligament* passes between the inferior pubic rami rounding the subpubic angle and strengthening the joint inferiorly.

Anterior to the interpubic disc the decussating fibres of rectus abdominis, external oblique and adductor longus strengthen the joint and provide additional anterior stability.

Movement: normally there is no movement; however, during pregnancy there is some separation (2 cm) at the symphysis pubis.

Palpation

The line of the symphysis pubis can be palpated anteriorly as a groove between the pubic bones where alignment of the joint can be checked.

MOVEMENTS

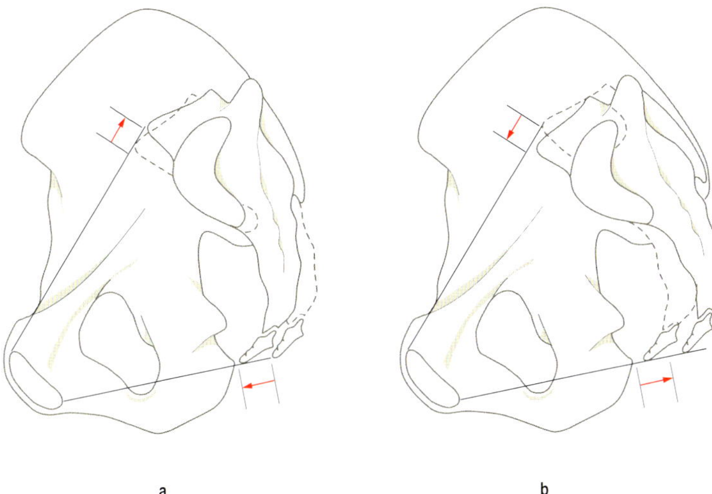

Movements of sacrum to increase the pelvic inlet (a) and outlet (b)

Given the need for stability during weight transference, little movement is possible due to the arrangement of articular surfaces and strong associated ligaments; nevertheless, slight gliding and rotatory movements are possible. When standing, compared to lying supine, the sacrum moves downwards 2 mm as well as undergoing 5° forward rotation.

Accessory movements: With the subject lying prone and the pelvis supported by the anterior superior iliac spines and pubic region a small rotation of sacrum against pelvis can be elicited when downward pressure is applied to the apex of sacrum.

During childbirth a complex movement of the sacrum involving rotation at the sacroiliac joints occurs. This is possible because of softening of sacroiliac and other ligaments during later stages of pregnancy, resulting in pelvic diameters increasing to facilitate passage of the foetal head. Initially, the sacral promontory moves posterosuperiorly to increase the anteroposterior diameter of the pelvic inlet by 3–13 mm. Once the foetal head has entered the pelvic canal, the sacral promontory moves inferoanteriorly to increase the anteroposterior diameter of the pelvic outlet by 15–18 mm.

Softening and laxity of ligaments when pregnant can cause instability and pain on weight-bearing, necessitating compression of pelvic joints by a support belt or even the use of walking aids. Sudden bending movements can also tear the posterior ligaments and dislocate joint surfaces, causing extreme pain during trunk flexion.

HIP

BONES

The hip joint is formed between the acetabulum of the pelvis and head of the femur.

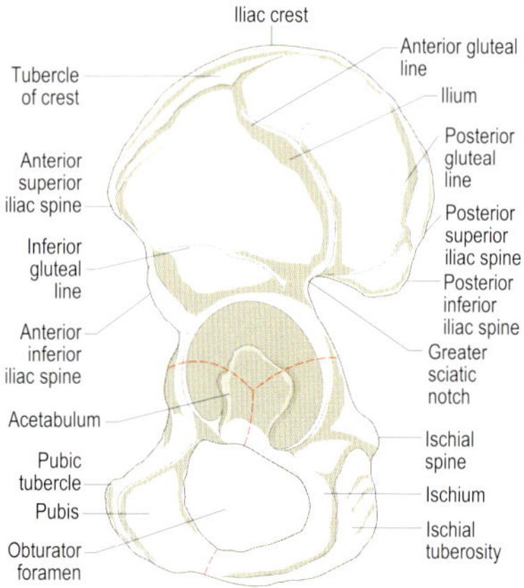

Lateral view left acetabulum

The acetabulum is the region of fusion of the ilium, pubis and ischium: the ilium contributes the superior two-fifths, the ischium the posteroinferior two-fifths and the pubis the anteroinferior one-fifth.

The acetabulum faces laterally, anteriorly and inferiorly, and the head of femur medially, anteriorly and superiorly.

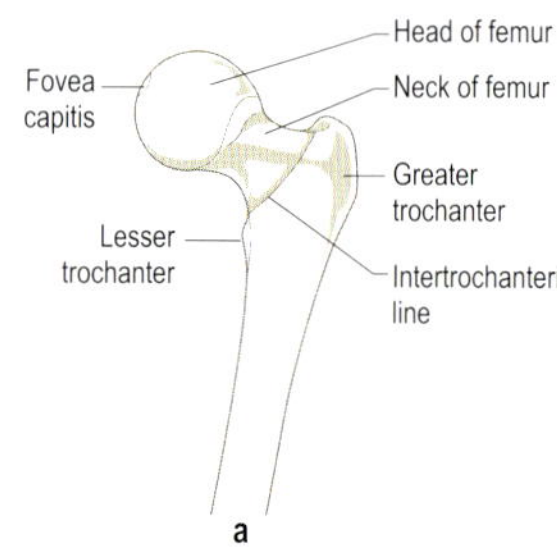

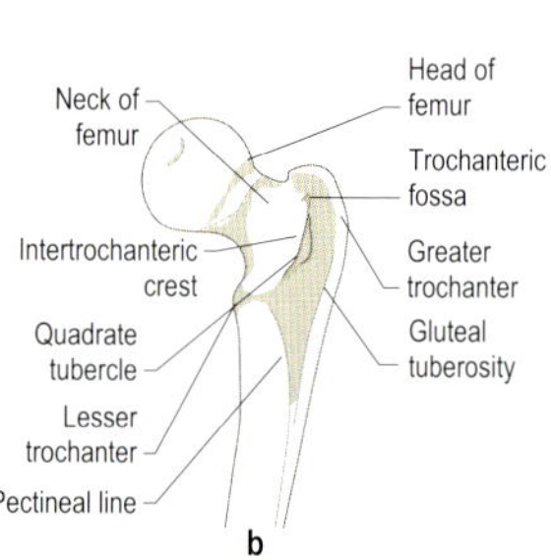

Proximal end of left femur: (a) anterior; (b) posterior

Palpation

The greater trochanter can be palpated 7–10 cm below the middle of the iliac crest, with the posterior border being palpable for 5 cm as it passes down to the shaft. The centre of each hip joint lies on a horizontal plane passing through the top of the greater trochanters 1 cm below the middle third of the inguinal ligament.

On anteroposterior radiographs of the pelvis a line drawn along the upper margin of the obturator foramen and inferior margin of the femoral neck to the medial side of the shaft describes a smooth curve: this is *Shenton's line*. The curve is not influenced by small changes in position, but is distorted in femoral fractures and dislocations.

AXES OF FEMUR

During embryonic development, and for 3 or 4 years after birth, the femoral shaft becomes adducted and medially rotated, resulting in the head and neck of femur becoming angled against the shaft in both frontal and horizontal planes.

In the frontal plane the angle between head, neck and shaft is the *angle of inclination*, which in adults is 125° (150° in the newborn).

In the horizontal plane outward rotation of head and neck against the shaft is the *angle of anteversion*, which in adults is approximately 10° (25° in infants and young children). If angles of inclination and anteversion are greater than 130° and 15° respectively, stability of the hip joint is decreased.

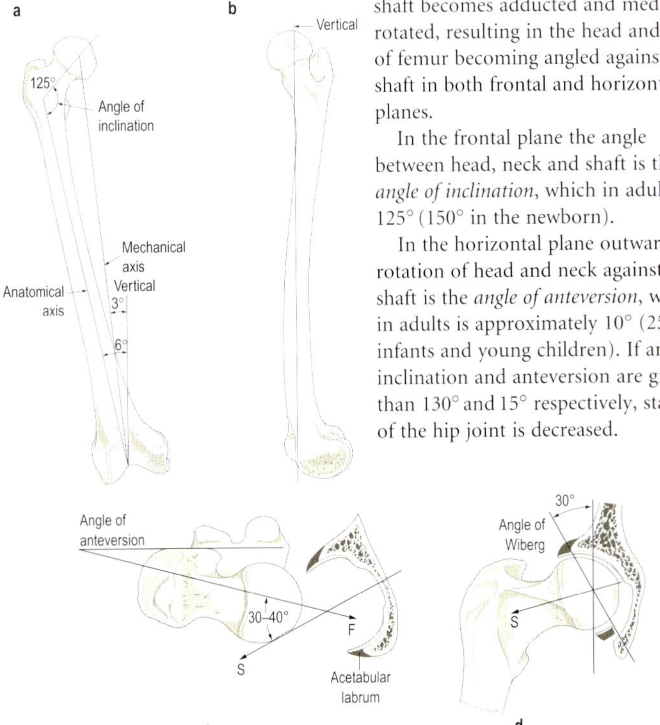

Axes of the femur: (a, d) anterior, (b) medial and (c) superior views

A line drawn between centres of the femoral head and knee joint is the *mechanical axis*, lying mainly outside the shaft due to length and angulation of the femoral neck. Medial and lateral rotation occur about this axis. The *mechanical axis* deviates 3° from vertical, while the *anatomical axis* of the shaft deviates 6° from vertical (as shown).

An angle of 30–40° exists between the axes of the acetabulum and femoral neck so that the anterior part of the femoral head articulates with the joint capsule. In addition, the lateral inferior inclination of the acetabulum forms an angle of 30–40° with the horizontal so that the superior part of the acetabulum overhangs the femoral head. An angle of 30° is formed between a vertical line through the centre of the head and a line drawn to the bony acetabular margin (*angle of Wiberg*): decreases in this angle reduce joint stability.

ARTICULAR SURFACES

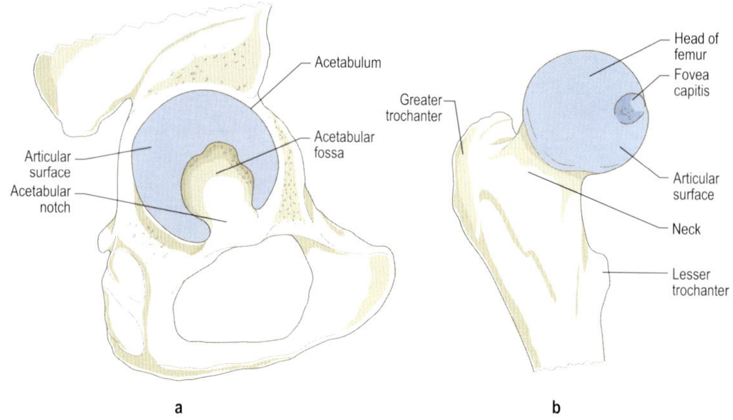

Articular surfaces of (a) acetabulum and (b) head of femur

Acetabulum Hemispherical socket deficient inferiorly (acetabular notch) on lateral surface of the innominate at the site of fusion of ilium, pubis and ischium: deepened by the fibrocartilaginous acetabular labrum. The semilunar articular surface, open below, is covered with hyaline cartilage: the central part of the acetabulum is the thin-walled non-articular acetabular fossa formed mainly by the ischium.

Head of femur Approximately two-thirds of a sphere, slightly compressed anteroposteriorly, covered by hyaline cartilage except for a small area superolaterally adjacent to the neck and at the fovea capitis.

Although reciprocally curved the articular surfaces are incongruent, resulting in limited surface area contact at low loads, increasing as load increases: this serves to distribute load and protect the underlying cartilage and bone from excessive stress.

Due to the relationship between femur and pelvis the superior surfaces of the femoral head and acetabulum sustain the greatest pressures and have the thickest cartilage. Only when the hip is weight-bearing and flexed does the anteromedial area of the acetabulum articulate with the inferior part of the femoral head.

CAPSULE AND SYNOVIAL MEMBRANE

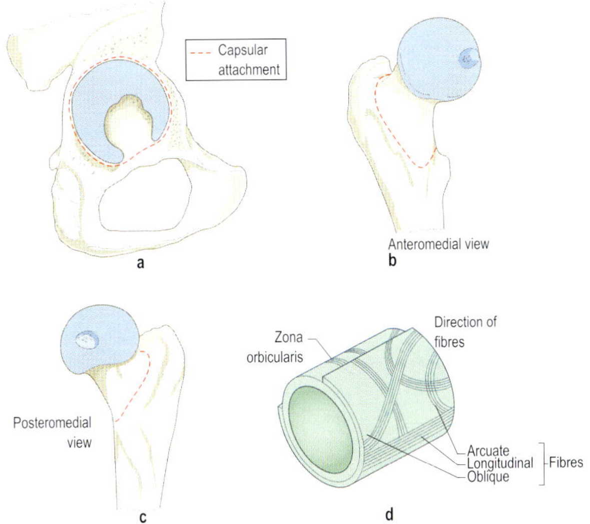

Attachments of hip joint capsule to (a) acetabulum and (b, c) femur; (d) capsular fibres

Joint capsule A strong fibrous capsule surrounds the hip joint, being thicker anteriorly and superiorly. It attaches to the acetabulum superiorly and posteriorly and to acetabulum and labrum elsewhere: at the acetabular notch the capsule attaches to the transverse ligament. Distally the capsule attaches to the intertrochanteric line and junction between the neck and trochanters of femur anteriorly: posteriorly the arched free border covers the medial two-thirds of the femoral neck. It is strengthened anteromedially by the reflected head of rectus femoris and laterally by gluteus minimus.

Longitudinal and *oblique fibres* pass from acetabulum to femur: *arcuate fibres* arch from one part of the acetabulum to another; deep *zona orbicularis fibres* have no bony attachments. On reaching the femoral neck some deeper longitudinal fibres turn upwards towards the articular margin as retinacular fibres and convey blood vessels to the head and neck.

Synovial membrane Lines and covers all non-articular surfaces, extending like a sleeve around the ligamentum teres (ligament of head) attaching to the margins of the fovea capitis. At the femoral attachment of the capsule the synovial membrane is reflected towards the head attaching to the articular margin. An extension of the membrane below the posterior capsule serves as a bursa for obturator externus. Anteriorly, the psoas bursa communicates with the joint space between the iliofemoral and pubofemoral ligaments.

INTRACAPSULAR STRUCTURES

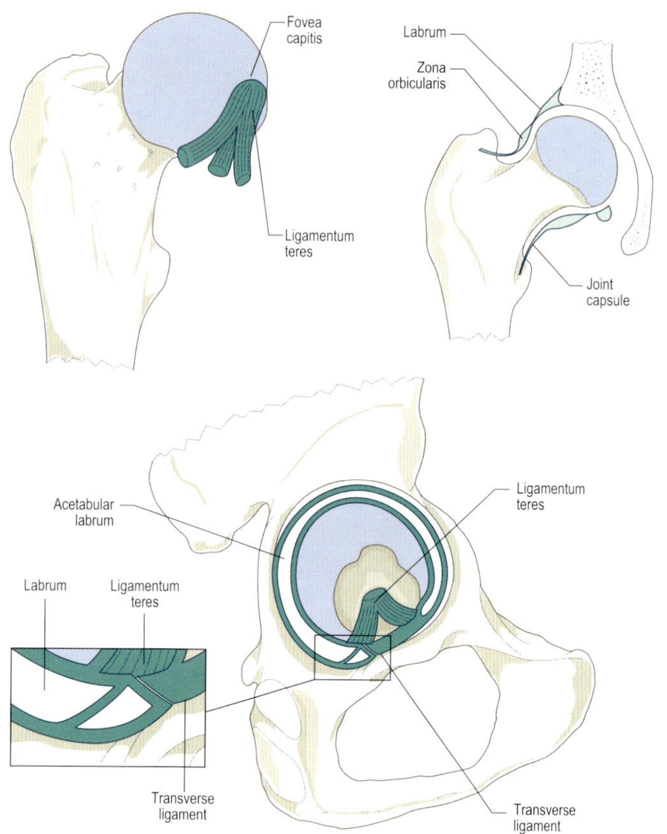

Attachments of intracapsular structures of the hip joint

Transverse acetabular ligament Strong bands of fibres bridging the acetabular notch, the superficial edge being level with the acetabular rim.

Acetabular labrum Triangular fibrocartilaginous ring attached to the acetabulum and transverse ligament. The apex is narrower than the maximum diameter of the femoral head which it cups.

Ligametum teres Weak flattened band of connective tissue between the margins of the acetabular notch and transverse ligament, and the fovea capitis. It is of little importance in adults; however, in children it conveys an artery (artery of the head of femur) supplying blood to the femoral head.

Acetabular fat pad Fibroelastic pad containing proprioceptive nerve endings lying within the acetabular fossa.

LIGAMENTS

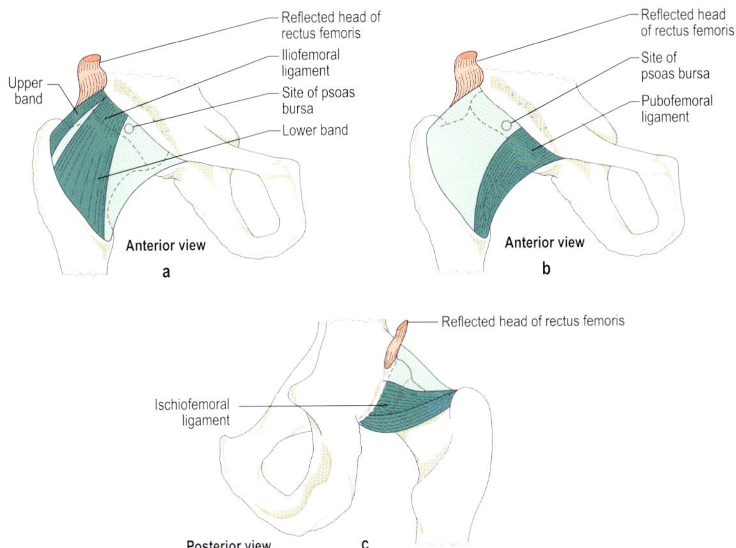

Capsular ligaments of the hip joint: (a) iliofemoral; (b) pubofemoral; (c) ischiofemoral

The main capsular longitudinal fibres form thickened bands named after their regional attachment around the acetabulum.

Iliofemoral ligament Thick strong triangular ligament anterior to the joint between the lower part of the anterior inferior iliac spine and adjacent acetabular rim, and intertrochanteric line: the outer bands are stronger than the thinner central part. It limits extension, lateral rotation, abduction (lower band) and adduction (upper band).

Pubofemoral ligament Strong narrow ligament anterior and inferior to the joint between the iliopubic eminence and superior pubic ramus, and the lower part of the intertrochanteric line. It limits extension, lateral rotation and abduction.

Ischiofemoral ligament Less well-defined spiral ligament posterior to the joint between the body of ischium, behind and below the acetabulum, and superior part of the neck and the root of greater trochanter. It limits extension, medial rotation and adduction.

MOVEMENTS

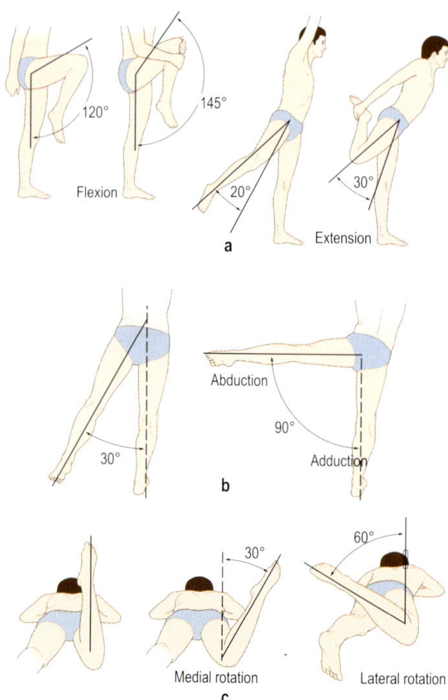

(a) Flexion and extension, (b) abduction and adduction, and (c) medial and lateral rotation of hip joint

Flexion and extension occur about a transverse axis, abduction and adduction about an anteroposterior axis, and medial and lateral rotation about the mechanical axis of femur: circumduction is also permitted. The axes of movement intersect at the centre of the femoral head. Because the head is at an angle to the shaft, all movements involve conjoint rotation of the head within the acetabulum.

Flexion is free but *extension* is limited by tension in the capsular ligaments.

Abduction and *adduction* are greatest with the hip flexed.

In *medial rotation* the shaft of femur moves anteriorly, while in *lateral rotation* it moves posteriorly. Rotation in both directions is freer when the hip is flexed. A small amount of hip rotation occurs automatically in the terminal phase of knee extension and at the beginning of knee flexion.

When assessing ranges of hip movement it is important to determine that there is no movement of the pelvis or vertebral column.

MUSCLES

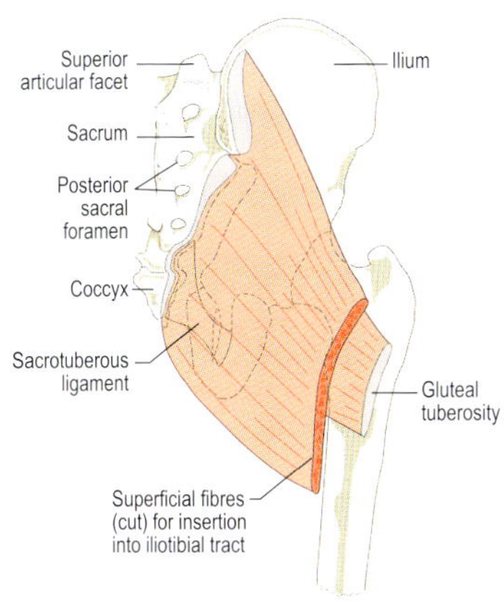

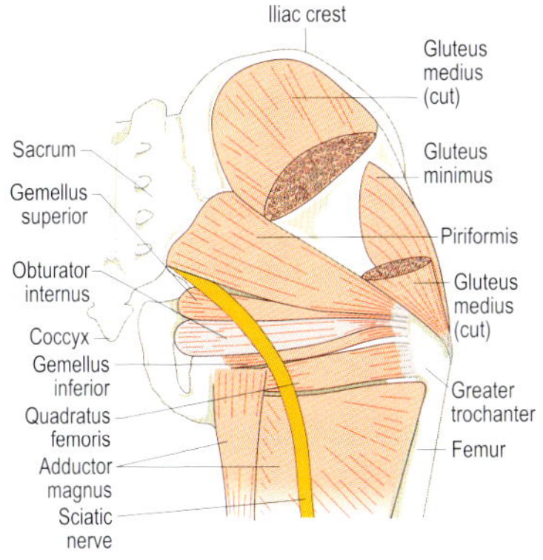

Posterior views of right hip

Gluteus maximus – extends, laterally rotates and assists abduction of hip.
Origin: gluteal surface and adjacent border of ilium, iliac crest, posterior aspect sacrum, side of coccyx, sacrotuberous ligament and fascia covering erector spinae.
Insertion: deep part to gluteal tuberosity of femur; superficial three-quarters to iliotibial tract.
Nerve supply: inferior gluteal nerve L5, S1, 2.

Gluteus maximus powerfully extends the flexed hip in actions such as running and climbing. Working with reversed origin/insertion it can raise the flexed trunk to an upright position. Through its attachment to the iliotibial tract it can help stabilize the knee.

Gluteus medius – abducts and medially rotates hip.
Origin: gluteal surface of ilium and covering fascia.
Insertion: superolateral side of greater trochanter.
Nerve supply: superior gluteal nerve L4, 5, S1.

Gluteus minimus – abducts and medially rotates hip.
Origin: gluteal surface of ilium deep and anterior to medius.
Insertion: anterosuperior aspect greater trochanter.
Nerve supply: superior gluteal nerve L4, 5, S1.

Gluteus medius and minimus (with tensor fascia lata) work with reversed origin/insertion to maintain the level of the pelvis in walking. When the right foot is raised these three muscles on the left contract to prevent the right side dropping. If weakened the result is a 'Trendelenburg gait'.

These muscles can be palpated by placing the hands with fingers pointing downwards over the anterior superior iliac spine and shifting weight from leg to leg.

MUSCLES

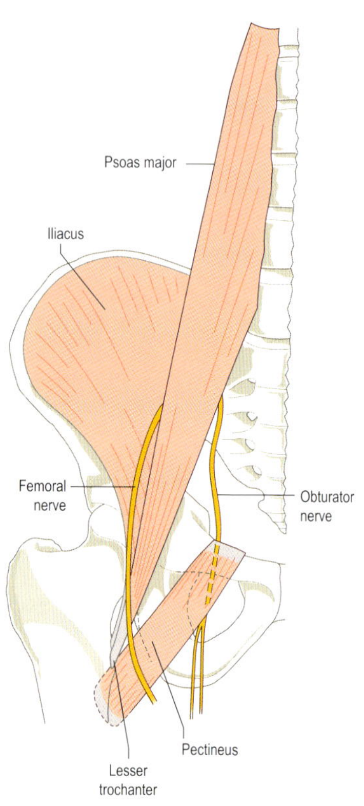

Anterior view right hip

Psoas major – flexes hip and laterally flexes trunk.
Origin: bodies of adjacent vertebrae and discs T12–L5; front of all lumbar transverse processes and tendinous arches over lumbar bodies.
Insertion: with iliacus into lesser trochanter of femur.
Nerve supply: anterior rami of L1, 2, 3 (4).

Iliacus – flexes hip.
Origin: upper and posterior two-thirds of iliac fossa, ala of sacrum and anterior sacroiliac ligament.
Insertion: into lesser trochanter of femur with psoas major.
Nerve supply: femoral nerve L2, 3.

The shared insertion of these two muscles means they are sometimes grouped together and called iliopsoas.

Pectineus – flexes and adducts hip.
Origin: superior ramus of pubis, iliopubic eminence and pubic tubercle.
Insertion: pectineal line on upper posterior part of femur.
Nerve supply: femoral nerve L2, 3 (occasionally by obturator nerve L3).

Working on both sides, psoas major is a strong trunk flexor as in sitting up from lying. This action tends to pull the lumbar spine forwards, increasing lordosis, which needs to be prevented by the action of other muscles. Double straight-leg raising when lying supine places considerable strain on the lumbar spine via the psoas attachment and is best avoided as an exercise.

Psoas major working on one side can laterally flex the trunk to the same side.

Iliacus is primarily a hip flexor but can assist psoas in actions such as sitting up from lying.

MUSCLES

Adductor magnus – adducts and extends hip.

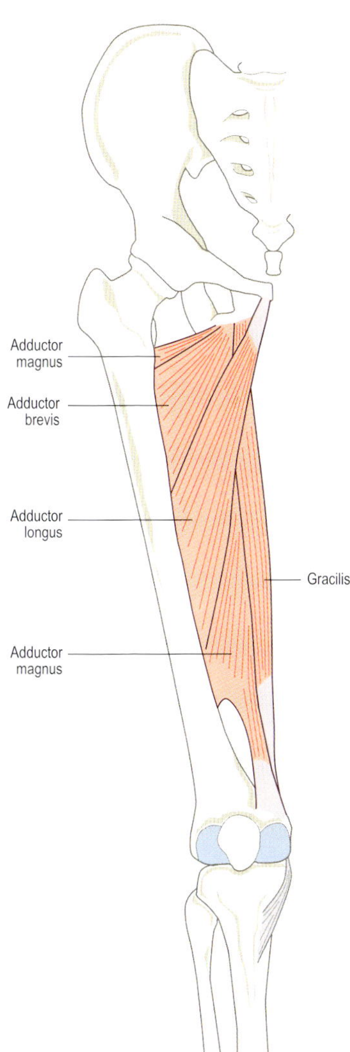

Anterior view right leg

Origin: femoral surface of ischiopubic ramus, inferior surface of ischial tuberosity.
Insertion: whole length of linea aspera and medial supracondylar ridge, with 'hamstring' part attaching to the adductor tubercle.
Nerve supply: obturator nerve L2, 3; 'hamstring' part – sciatic nerve L4.

Adductor longus – adducts hip.
Origin: anterior aspect body of pubis.
Insertion: middle half linea aspera of femur.
Nerve supply: obturator nerve L2, 3, 4.

Adductor brevis – adducts hip.
Origin: body and inferior ramus of pubis.
Insertion: upper half linea aspera of femur.
Nerve supply: obturator nerve L2, 3, 4.

Gracilis – adductor of hip and flexor of knee.
Origin: body and inferior ramus of pubis.
Insertion: medial surface shaft of tibia between sartorius and semitendinosus.
Nerve supply: obturator nerve L2, 3.

Adduction can be a very strong movement but is rarely performed on its own. Functionally, the adductors work with other muscles around the hip during gait, e.g. adductors on the left side contract to move body weight over the left foot, allowing the right (non-weight-bearing) leg to step forward. Hip abductors on the left side work to maintain a level pelvis during this phase of gait, i.e. hip abductors and adductors on the *same* side work together.

The adductors are well developed in speed skaters and horse riders.

The strength of adductors can cause problems in situations of increased muscle tone where surgical or pharmaceutical intervention may be required to allow more normal function.

MUSCLES

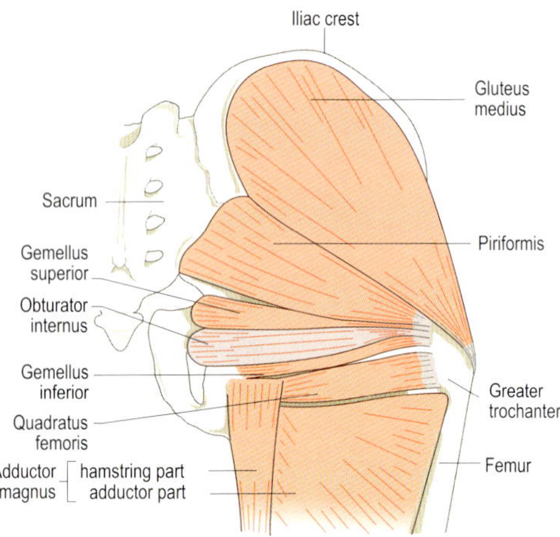

Posterior view right hip

The muscles described below are usually called the 'lateral rotators' as they all have common actions of lateral rotation and stabilization of the hip.

Piriformis

Origin: front of 2nd, 3rd, 4th sacral segments lateral to sacral foramina, gluteal surface ilium and sacrotuberous ligament.
Insertion: upper border and medial side greater trochanter of femur.
Nerve supply: anterior rami sacral plexus L5, S1, 2.

Quadratus femoris

Origin: ischial tuberosity just below acetabulum.
Insertion: quadrate tubercle on intertrochanteric crest.
Nerve supply: nerve to quadratus femoris L4, 5, S1.

Obturator internus

Origin: internal surface obturator membrane and surrounding bony margin.
Insertion: medial surface greater trochanter of femur.
Nerve supply: nerve to obturator internus L5, S1, 2.

Gemellus superior and inferior

Origin: superior – ischial spine; inferior – upper part ischial tuberosity.
Insertion: both muscles blend to insert with tendon of obturator internus into greater trochanter of femur.
Nerve supply: superior – nerve to obturator internus; inferior – nerve to quadratus femoris.

Obturator externus (not shown)

Origin: outer surface obturator membrane and surrounding bone.
Insertion: trochanteric fossa of femur.
Nerve supply: obturator nerve L3, 4.

Functionally, these muscles hold the head of femur in the acetabulum, acting as 'contractile ligaments'. Complex movements such as transferring weight in the sitting position, e.g. chair to wheelchair, utilize medial and lateral rotation when the feet are fixed on the ground.

MUSCLES

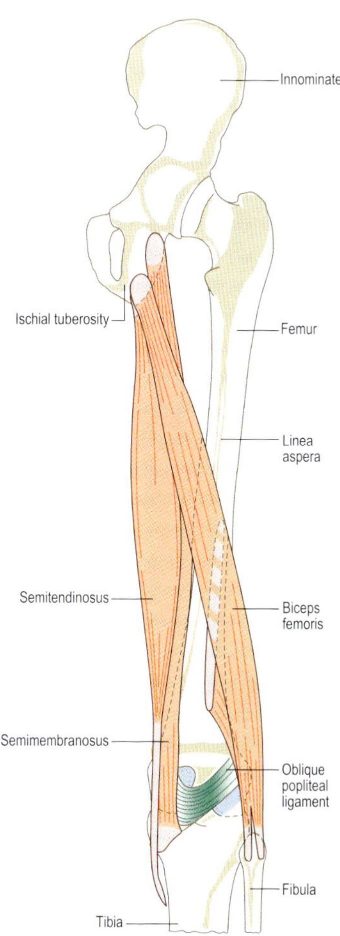

Posterior view right thigh

Hamstrings – as a group extend hip and flex knee.

Semitendinosus

Origin: lower medial facet on ischial tuberosity.

Insertion: medial surface upper part of tibia (behind sartorius and gracilis).

Nerve supply: sciatic nerve L5, S1, 2.

Semimembranosus

Origin: upper lateral facet on ischial tuberosity.

Insertion: posteromedial surface of medial tibial condyle.

Nerve supply: sciatic nerve L5, S1, 2.

Biceps femoris

Origin: long head – lower medial facet on ischial tuberosity; short head – lower half lateral lip linea aspera of femur.

Insertion: head of fibula.

Nerve supply: sciatic nerve L5, S1, 2.

- Flexion of knee is their most important action as other large muscles can extend hip.
- Hamstrings work strongly to raise the flexed trunk to the upright position.
- Balancing the pelvis on the femurs anteroposteriorly is a coordinated action between abdominals anteriorly and gluteus maximus and hamstrings posteriorly. This pelvic tilting has a significant effect on lumbar lordosis.
- Hamstrings decelerate forward movement of tibia during the swing phase of gait to prevent the knee snapping into extension.
- Hamstrings are commonly injured during sprinting probably because they are contracting strongly and working over two joints simultaneously, i.e. raising trunk from the start position whilst the hip is being flexed.

RELATIONS

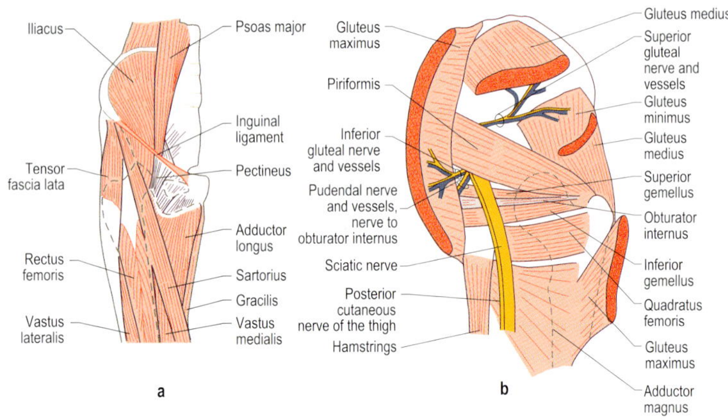

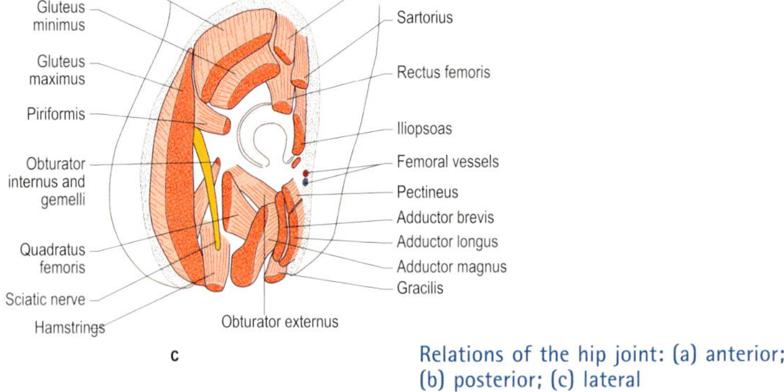

Relations of the hip joint: (a) anterior; (b) posterior; (c) lateral

Muscles surround the anterior, posterior and lateral aspects of the hip.

Anteriorly, the tendons of iliacus and psoas and the femoral nerve overlie the joint capsule with the femoral artery and vein lying on psoas and pectineus. *Posteriorly*, the nerve to quadratus femoris lies directly on the joint capsule, while the sciatic nerve is separated from it by obturator internus and the gemelli.

The *fascia lata* (deep fascia of thigh) invests the soft tissues of the lower limb: an opening anteriorly below the level of the hip joint transmits a number of superficial arteries, the long saphenous vein and efferent lymphatic vessels from superficial lymph nodes. Enlargement of the superficial or deep inguinal nodes will present as a swelling in the groin.

APPLIED ANATOMY

Dislocation of hip Posterior dislocation is the most common, usually the result of trauma producing forces longitudinal to the femur in a sitting position, e.g. driving or front-seat passenger position in a road traffic accident.

Developmental dysplasia of hip Formerly known as congenital dislocation, this covers a wide range of disorders stemming from a developmental abnormality, e.g. dislocation or partial displacement during the neonatal period, often resulting from general joint laxity and an unusually shallow acetabulum. Development of the hip joint may be delayed and changes in patterns of stress applied to the hip can lead to remodelling of trabeculae in the femoral head and neck, causing deformation of the bone. This changes the relationship between head and neck of femur and shaft, leading to *coxa valga* or *coxa vara*.

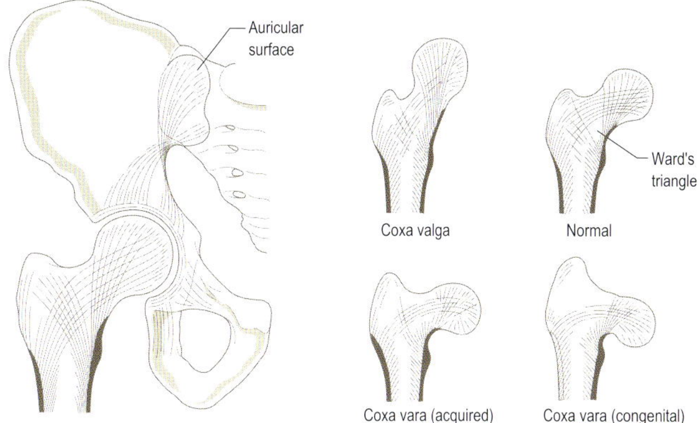

(a) Arrangement of bony trabeculae in innominate and upper part of femur; (b) changes in normal Ward's triangle in coxa valga and coxa vara showing internal remodelling

Fracture of neck of femur The neck of femur is a common site of fracture, particularly in elderly females. Factors contributing to this include a smaller head of femur in the female, thus increasing stress across the joint. The wider female pelvis also alters leverage and requires greater abduction forces to control pelvic tilt. Female hormonal changes leading to reduction in bone density also play a part. Following fracture the vascular supply to the head of femur may be reduced leading to *avascular necrosis*, which is why the head and neck are frequently replaced with a prosthesis (*hemiarthroplasty*). **NB**: In *osteoarthritis* of the hip both articular components may be replaced surgically (*arthroplasty – total hip replacement*).

Perthes' disease may also result from disrupted blood supply to the head of femur in early childhood. Initial necrosis of the head is followed by revascularization and distorted growth, resulting in a typically 'flattened' head of femur (*coxa plana*) or enlargement (*coxa magna*).

KNEE

INTRODUCTION

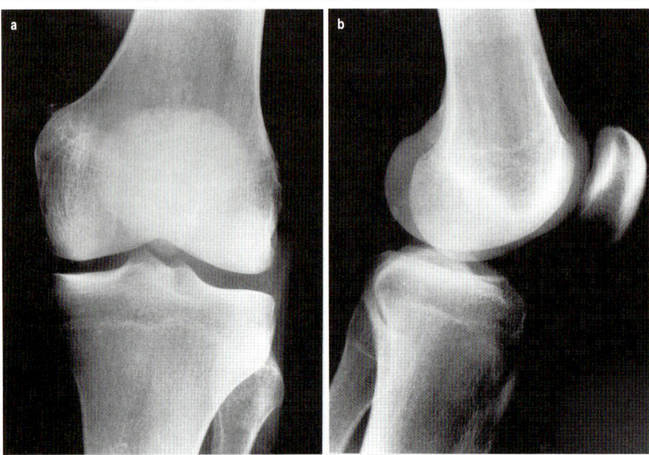

(a) Anteroposterior and (b) lateral radiographs of knee

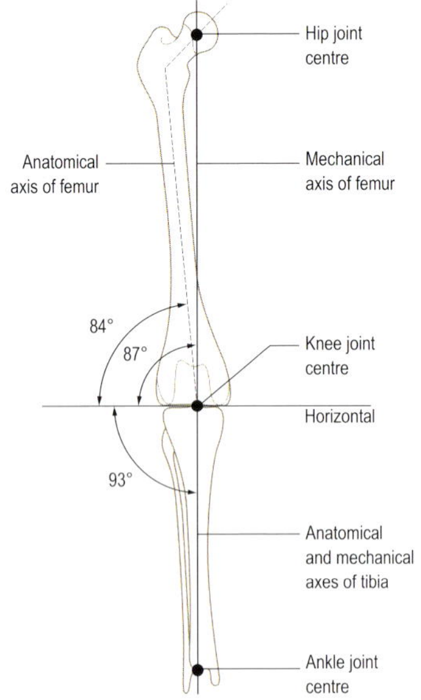

Anatomical and mechanical axes of femur and tibia

Large complex bicondylar hinge joint between femur, tibia and patella providing both mobility and stability. By lengthening and shortening the lower limb the knee is important in locomotion. With the ankle joint it provides strong forward propulsion of the body.

The anatomical axes of the femur and tibia form an outward angle of 170–175° (*femorotibial angle*). Increases (*genu varus* – 'bowlegs') or decreases (*genu valgus* – 'knock knees') in this angle occur in some pathological conditions. Genu valgus is not uncommon in toddlers, generally disappearing with growth.

The transverse axis of the knee is horizontal in the frontal plane: it does not bisect the femorotibial angle so that the angle between it and the tibia is larger than that with the femur.

BONES

The knee joint is formed between the distal end of femur, proximal end of tibia and posterior surface of patella.

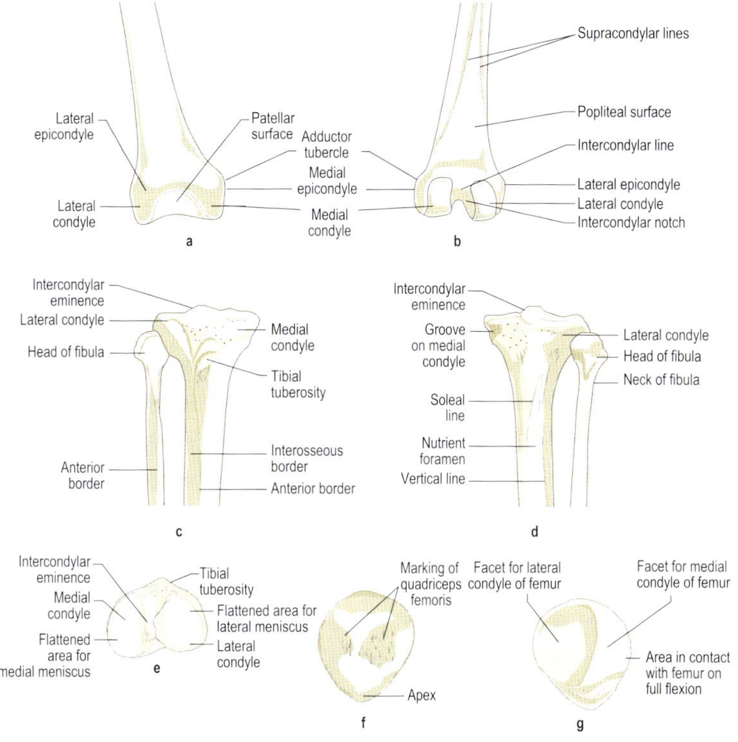

Distal end of right femur (a, anterior; b, posterior; e, superior) and proximal end of right tibia (c, anterior; d, posterior); left patella (f, anterior; g, posterior)

> **Palpation**
>
> *Femur:* Articular margins of the medial and lateral condyles can be palpated, as can the prominent epicondyles projecting from the outer surfaces of each condyle. The adductor tubercle can be palpated above the medial condyle.
>
> *Tibia:* Anterior, medial and lateral margins of the tibial condyles can be palpated. The tibial tuberosity can be felt and seen at the upper end of the anterior border of tibia 2 cm below the condylar margin.
>
> *Patella:* The whole margin and anterior surface can be palpated.
>
> *Joint line:* The joint line can be palpated medially and laterally between femoral and tibial condyles and followed anteriorly.

ARTICULAR SURFACES

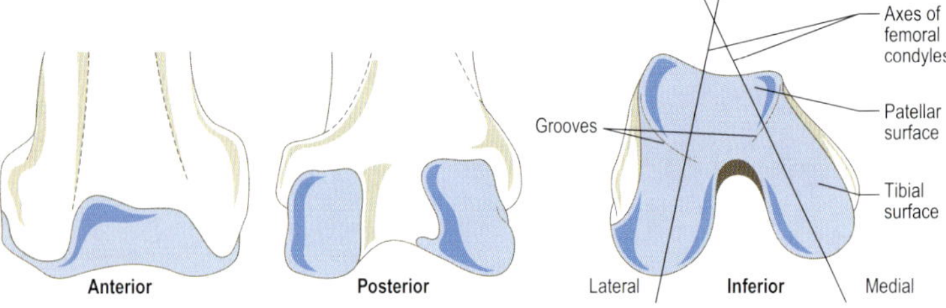

Articular surfaces of right femur

Femur The convex femoral condyles, longer anteroposteriorly than transversely, diverge posteriorly, with the medial condyle being narrower and jutting out more than the lateral. The intercondylar notch continues the groove of the patella surface. Faint grooves separate condylar and patellar surfaces, with the patellar surface divided by a well-marked groove into smaller medial and larger more prominent lateral parts.

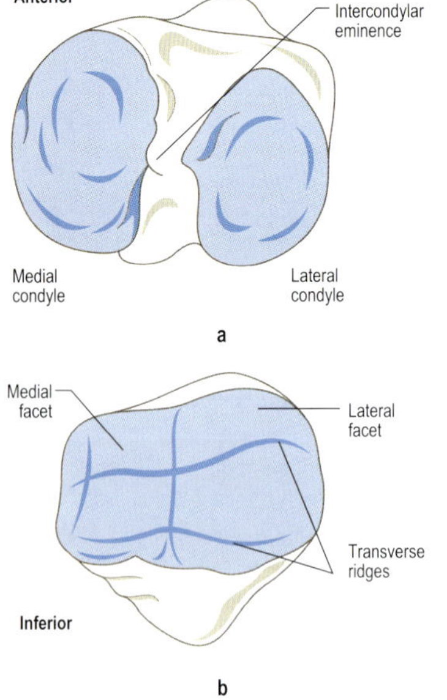

Articular surfaces of (a) right tibia and (b) right patella

Tibia Relatively flat articular surfaces separated by intercondylar eminence with triangular areas anterior and posterior, the eminence lodging in the intercondylar notch of femur. The oval concave medial articular surface is larger than the rounded lateral surface, concave transversely but concavoconvex anteroposteriorly.

Patella Oval articular surface divided into larger lateral and smaller medial areas by a vertical ridge. Another faint ridge separates a medial perpendicular facet from the main medial area.

CAPSULE AND SYNOVIAL MEMBRANE

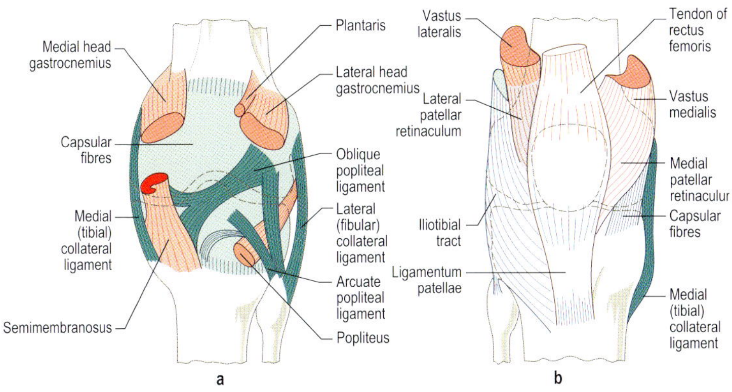

Knee joint capsule, right side: (a) posterior; (b) anterior

Joint capsule The knee is surrounded by a thick ligamentous sheath composed mainly of muscle tendons and their expansions: there is no complete independent fibrous capsule. Anteriorly, the capsular attachment to the femur is deficient, blending with the quadriceps tendons: its attachment to the tibia is more complete, being deficient only in the region of the tibial tuberosity. Posteriorly, true capsular fibres pass vertically from above the articular surface to the posterior border of tibia, being strengthened by the *oblique popliteal ligament* (an expansion of semimembranosus tendon). At the sides capsular fibres pass from femoral to tibial condyles, blending posteriorly with a ligamentous network and anteriorly with tendinous expansions of quadriceps femoris: the lower lateral capsule is strengthened by the *arcuate popliteal ligament* from the fibular head.

Synovial membrane Lines the joint capsule attaching to articular margins of femur, tibia and patella. On the tibia it is reflected forward around the cruciate ligaments so that they are intracapsular but extrasynovial. Above the patella between the femoral shaft and quadriceps femoris the synovial membrane extends 6 cm as the suprapatellar bursa, to which are attached a few muscle fibres from vastus intermedius (articularis genu). Synovial recesses extend behind each femoral condyle and deep to the popliteus tendon. Bursae associated with tendons crossing the joint do not communicate directly with the joint space.

COLLATERAL LIGAMENTS

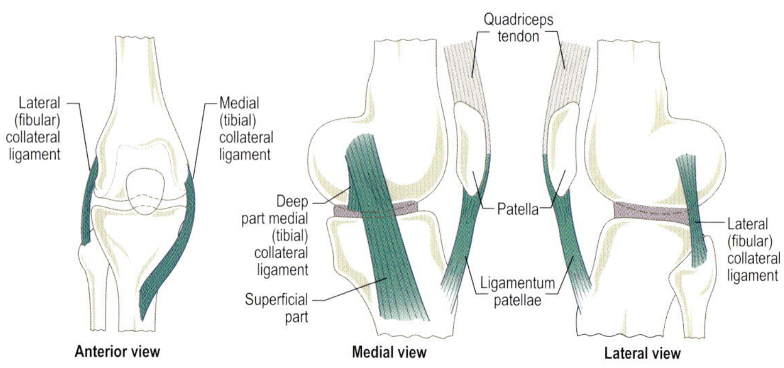

Medial and lateral collateral ligaments of the knee joint

Medial (tibial) collateral ligament Strong band passing downward and forward from the medial epicondyle of femur to the medial condyle and shaft of tibia. The most superficial fibres extend below the level of the tibial tuberosity, deeper fibres pass from femur to tibia, while the deepest fibres spread out to attach via the capsule to the medial meniscus.

Lateral (fibular) collateral ligament Round cord, 5 cm long, passing downward and backward from the lateral epicondyle of femur to the lateral surface of the fibular head anterior to the apex: it does not blend with the joint capsule.

> Collateral ligaments provide mediolateral stability at the knee: in the extended or hyperextended knee lateral displacement of the tibia indicates disruption of the medial collateral ligament; similarly, medial displacement of the tibia indicates disruption of the lateral collateral ligament. The direction of the collateral ligaments is such that they tighten during extension and contribute to the 'locking' mechanism of the knee.

CRUCIATE LIGAMENTS

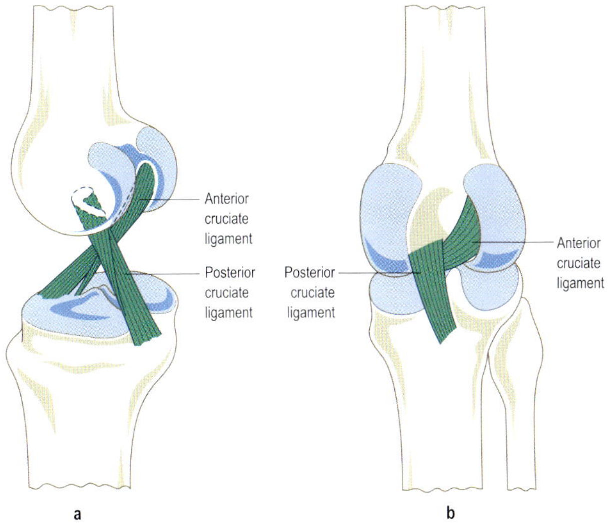

The *anterior cruciate ligament* (*ACL*) passes posteriorly, laterally and proximally from the anterior tibial spine to the medial surface of the lateral femoral condyle, spiralling medially through 110° as it does so. The anteromedial band limits flexion and the posterolateral band limits extension.

The *posterior cruciate ligament* (*PCL*) passes anteriorly, medially and proximally from the posterior intercondylar area of the tibia to the lateral side of the medial femoral condyle crossing the ACL on its medial aspect. The anterolateral band limits flexion and the posteromedial limits extension.

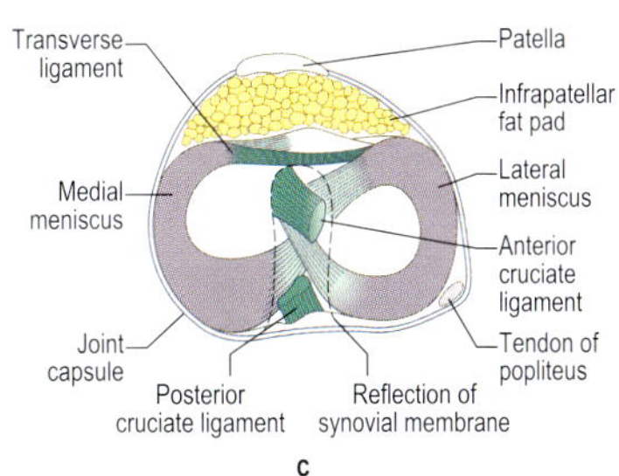

Cruciate ligaments, right side: (a) medial, (b) posterior and (c) transverse views

The cruciate ligaments provide resistance to anterior and posterior displacement of the tibia with respect to the femur: the ACL 86% of anterior restraint, the PCL 94% of posterior restraint. They also provide some mediolateral stability: the ACL 30% of medial tibial displacement, the PCL 36% of lateral tibial displacement.

MENISCI

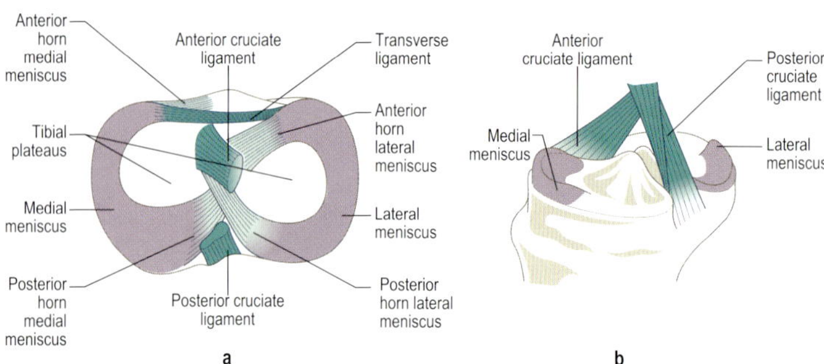

Menisci of the right knee: (a) superior and (b) oblique views

The menisci are intra-articular crescentic-shaped fibrocartilaginous structures triangular in cross-section interposed between femoral and tibial condyles. The periphery is attached to the joint capsule, the medial meniscus being firmly anchored to the medial collateral ligament.

The superior surface is smooth and concave: the inner free border is thin. The *medial meniscus* is broader posteriorly than anteriorly; the *lateral meniscus* is of uniform width.

Anterior and posterior horns attach the menisci to the intercondylar eminence.

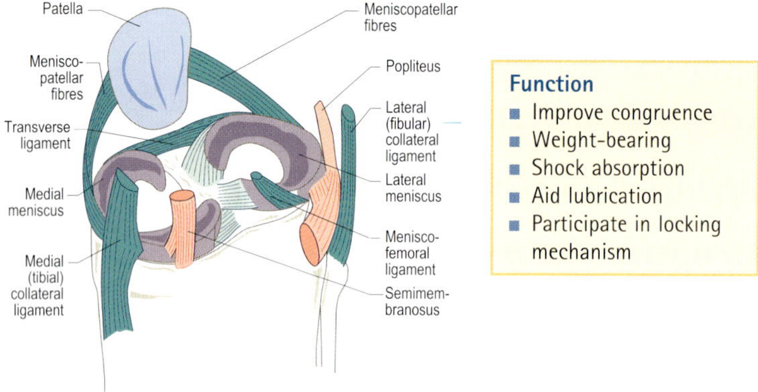

Attachments of menisci of the right knee

Function
- Improve congruence
- Weight-bearing
- Shock absorption
- Aid lubrication
- Participate in locking mechanism

Posteriorly the medial meniscus is attached to the oblique popliteal ligament via the joint capsule and the lateral meniscus to the popliteus tendon: the lateral meniscus usually contributes a slip to the PCL (meniscofemoral ligament). From their outer margins thickenings of the joint capsule attach the menisci to the sides of the patella (meniscopatellar fibres).

FLEXION AND EXTENSION

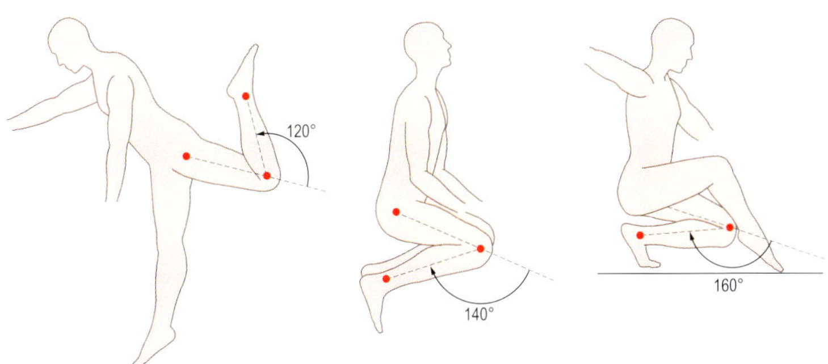

Flexion of the knee joint

The range of flexion depends on hip position and whether movement is active or passive. Some movement, usually passive, of the tibia beyond alignment of the long axes of the thigh and leg (*hyperextension*) may be possible.

Flexion is normally limited by contact of thigh and calf muscles; however, if movement is arrested before then it could be due to retraction of quadriceps or shortening of capsular ligaments.

In flexion the menisci move posteriorly on the tibial condyles so their posterior parts project beyond the tibial condyles: in extension the menisci move anteriorly with their anterior parts projecting beyond the anterior tibial condyles.

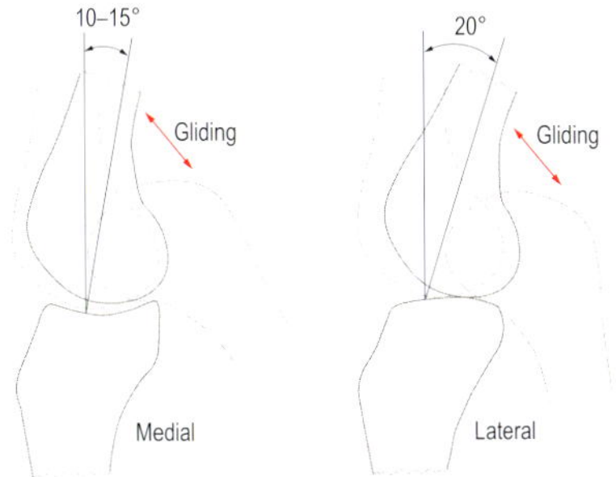

Rolling and gliding of femoral and tibial condyles during flexion and extension: - - - fully extended; — — fully flexed; —— limit of pure rolling

Flexion involves a combination of rolling and gliding of femoral condyles on tibial condyles. Flexing from full extension the femoral condyles begin to roll without gliding but by the end of range they glide without rolling. The extent of rolling and gliding differs for medial and lateral femoral condyles.

The change from rolling to gliding is significant for knee function where both mobility and stability are needed. The initial 15–20° of pure rolling corresponds to the range of knee flexion during the support phase of walking, when stability is the prime requirement.

MEDIAL AND LATERAL ROTATION

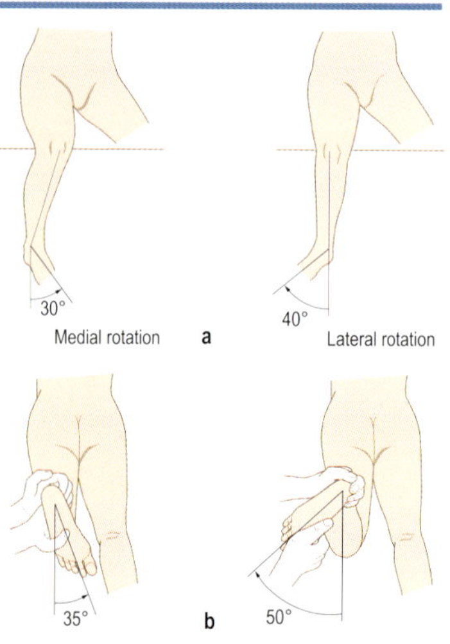

Rotation is only possible with the knee flexed. In lateral rotation of tibia on femur the lateral femoral condyle moves forwards and the medial condyle backwards over the tibial condyle: the reverse occurs in medial rotation. The menisci follow the movements of the femoral condyles.

Automatic rotation of the knee is associated with the terminal part of extension and the beginning of flexion. With the tibia fixed the femur medially rotates at the end of extension and laterally rotates at the start of flexion. With the femur fixed the tibia undergoes lateral rotation at the end of extension and medial rotation at the beginning of flexion. This is 'locking' and 'unlocking' of the knee.

Axial rotation at the knee joint: (a) active; (b) passive

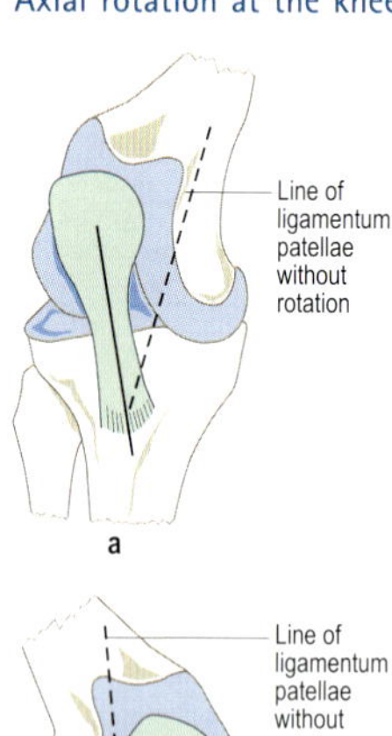

Patellar movements during axial rotation: (a) medial; (b) lateral

During rotation the patella moves in a frontal plane with respect to the tibia. In medial tibial rotation the patella is dragged laterally so that the ligamentum patellae runs obliquely inferiorly and medially: the opposite occurs in lateral tibial rotation.

Accessory movements: When fully extended no accessory movements are possible. When flexed to 25° the tibia can be moved anteriorly and posteriorly by applying appropriate force: it can also be rocked medially and laterally. Applying longitudinal force to the leg the tibia can be distracted from the femur.

MUSCLES

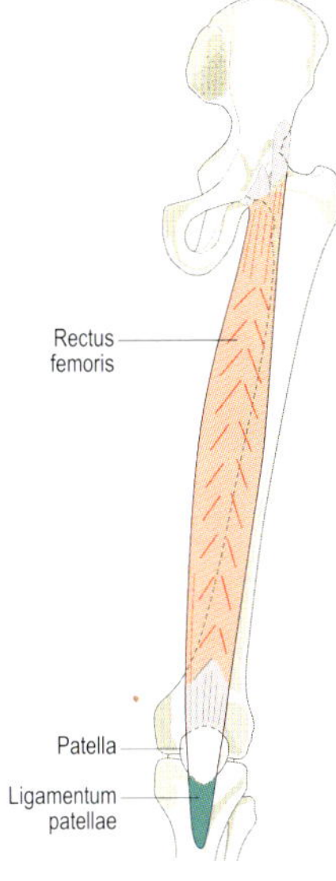

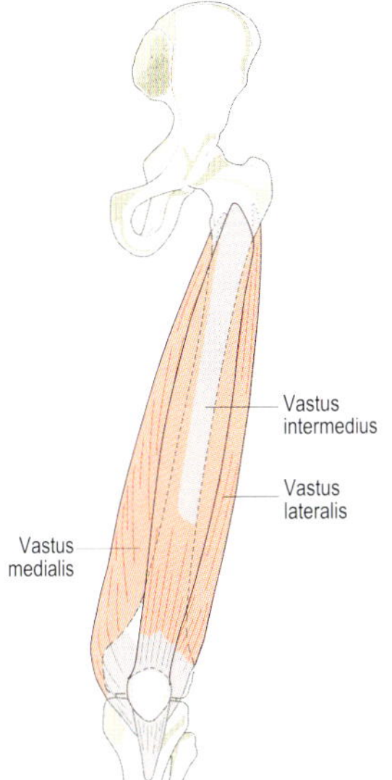

Anterior views left thigh

Quadriceps femoris – all extend the knee; rectus femoris can also flex the hip.

Rectus femoris

Origin: straight head – from anterior inferior iliac spine; reflected head – from above acetabulum.

Insertion: base of patella.

Vastus lateralis

Origin: upper part intertrochanteric line, lower border greater trochanter, lateral side gluteal tuberosity, upper half lateral lip linea aspera, lateral intermuscular septum.

Insertion: base and lateral border patella.

Vastus medialis

Origin: lower part intertrochanteric line, spiral line, medial lip linea aspera, upper part medial supracondylar line, medial intermuscular septum.

Insertion: medial border patella.

Vastus intermedius

Origin: upper two-thirds anterior and lateral surfaces of femur.

Insertion: base of patella with rectus femoris.

Ligamentum patellae

All four heads of quadriceps contribute to formation of the ligamentum patellae, running from apex of patella to tibial tuberosity, acting as tendon of insertion of quadriceps.

Nerve supply: all supplied by femoral nerve L2, 3, 4.

> Quadriceps is the powerful extensor of the knee, but frequently works eccentrically to control flexion produced by gravity, e.g. sitting, stepping down, squats.
>
> Vastus medialis has horizontal fibres which prevent lateral dislocation of the patella and oblique fibres (vastus medialis obliquus, VMO) which help fully extend the knee.
>
> Rectus femoris is bipennate and via attachment to the pelvis is also a hip flexor. This needs to be considered when stretching quadriceps as a group, i.e. hip should be extended whilst knee is flexed.

MUSCLES

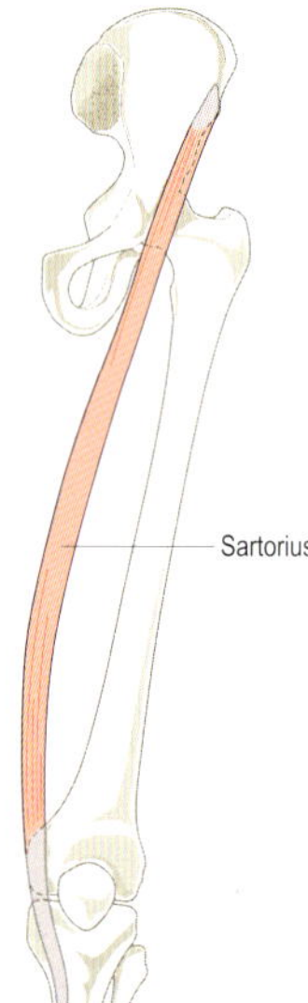

Anterior view left thigh

Sartorius – flexes, laterally rotates and abducts hip; flexes knee.
Origin: anterior superior iliac spine.
Insertion: upper medial part shaft of tibia in front of semitendinosus and gracilis.
Nerve supply: femoral nerve L2, 3.

> Sartorius is a strap muscle having the longest muscle fibres in the body. It is called the 'tailor's muscle', as combined action at hip and knee places heel on the opposite thigh in sitting.

Tensor fascia lata – abducts and medially rotates hip; also assists extension of knee.
Origin: anterior superior iliac spine and adjacent iliac crest.
Insertion: between the two layers of the iliotibial tract which attaches to the lateral tibial condyle.
Nerve supply: superior gluteal nerve L4, 5.

> Functionally, tensor fascia lata is a hip abductor, lying next to gluteus medius and minimus, sharing the same nerve supply. Via attachment to the iliotibial tract it helps knee extension.

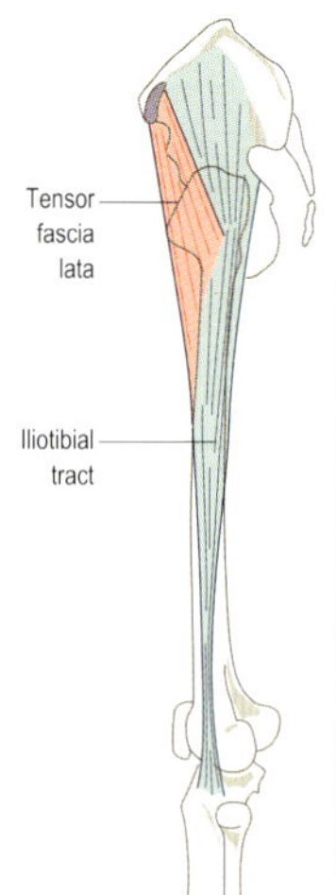

Lateral view left thigh

Popliteus – posterior to knee joint: flexes knee and medially rotates tibia.
Origin: outer surface lateral condyle of femur.
Insertion: posterior surface of tibia above soleal line.
Nerve supply: tibial nerve L5.

> Popliteus laterally rotates the femur on tibia to unlock knee when foot is fixed; also pulls the lateral meniscus backwards, preventing its entrapment.

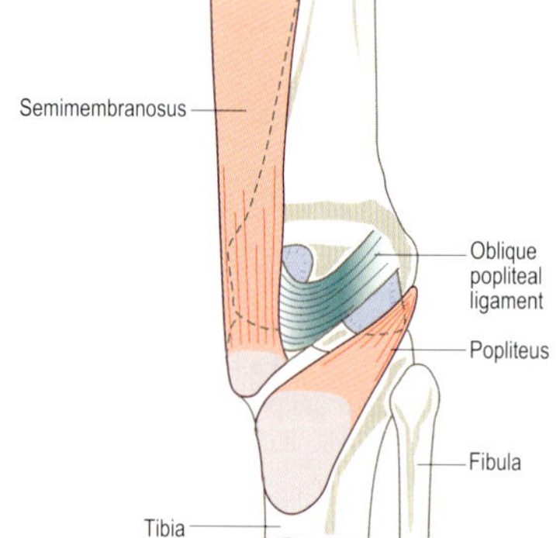

Posterior view right knee

RELATIONS

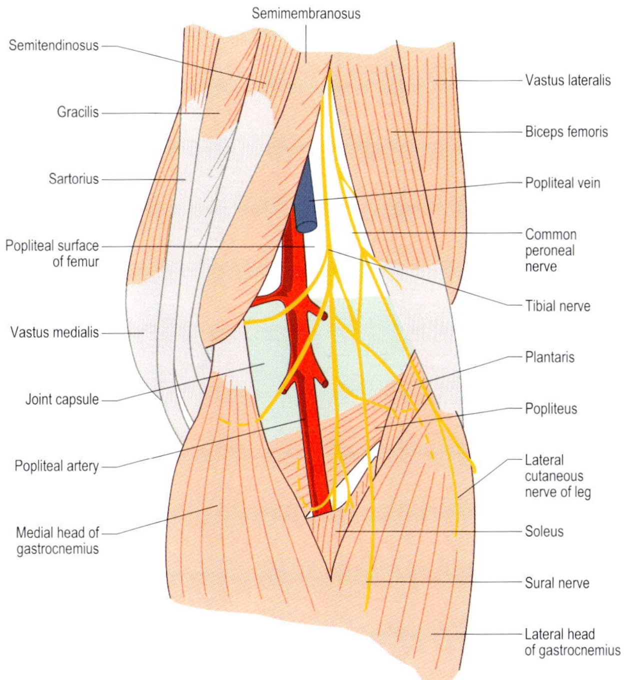

Popliteal fossa and posterior relations of right knee

Posterior to the knee joint is the diamond-shaped popliteal fossa, a transitional region between thigh and leg, with its boundaries formed by muscles. The floor is formed from above down by the popliteal surface of femur, posterior part of the knee joint capsule (reinforced by the oblique popliteal ligament) and popliteus. The fossa is covered by dense popliteal fascia continuous with that of the thigh and leg.

Directly below the fascia are the tibial and common peroneal divisions of the sciatic nerve, the tibial nerve passing vertically through the fossa while the common peroneal nerve passes under cover of biceps femoris. The popliteal vessels enter superomedially through the adductor hiatus, with the artery lying directly on the floor. The artery gives genicular branches supplying the knee joint before dividing into anterior and posterior tibial arteries at the level of the tibial tuberosity. The popliteal vein receives the small saphenous vein. The artery and vein are embedded in fat and areolar connective tissue in which popliteal lymph nodes are found.

APPLIED ANATOMY

Lesions of menisci

Tears are common in young adults, usually occurring when weight is taken on the flexed knee with added rotation at the joint – hence the high incidence in footballers. The meniscus may be torn along its length by the grinding forces between femur and tibia. The medial meniscus is most commonly torn as it is less mobile than the lateral due to its attachment to the tibial collateral ligament. With increasing age tears can occur with relatively little force as fibrosis restricts mobility of the menisci.

Varus and valgus deformities

In adults these are likely to be secondary to disorders such as rheumatoid arthritis (usually valgus) or osteoarthritis (usually varus) where the medial or lateral compartment of the knee is most affected, leading to a loss of joint space and collapse on the same side. In children such deformities are often considered normal stages of development and most correct spontaneously by puberty.

Patellofemoral disorders

Disorders of the patella can give rise to anterior knee pain and may result from maltracking or instability of the patella. Repeated dislocation, usually lateral, can damage the surface of the femoral condyle and lead to secondary osteoarthritis. Chondromalacia patella is a softening or fibrillation of the articular surface of patella, often resulting from the above mechanical knee problems.

Osteochondritis dessicans

This presents as loose bodies in the knee resulting from fragments of cartilage or underlying bone separating from the femoral condyles, either from trauma or repeated impact with the patella or adjacent tibial ridge. The fragments can cause 'locking' or 'giving way' of the knee.

Osgood–Schlatter's disease

This is a traction injury of the epiphysis at the tibial tuberosity where the patellar ligament inserts, causing it to become prominent and painful.

Bursitis

There are a number of bursae around the knee that can become inflamed and present with a fluctuating swelling. They include *prepatellar bursitis* ('housemaid's knee'), *infrapatellar bursitis* ('clergyman's knee') anterior to the knee joint: *semimembranosus bursa* (enlarged but usually painless), bulging of the posterior capsule and synovial pouch (*popliteal cyst*) posterior to the joint. The latter is sometimes called a 'Baker's cyst' and is usually associated with osteoarthritis or rheumatoid arthritis.

Ligament disorders

The *collateral ligaments* can be partially torn or even ruptured by a medial or a lateral force directed to the knee.

The *cruciate ligaments* can similarly be damaged by sudden anterior or posterior forces of femur on tibia. Because the cruciates are vital to knee stability they are frequently replaced surgically when ruptured.

TIBIOFIBULAR JOINTS

BONES

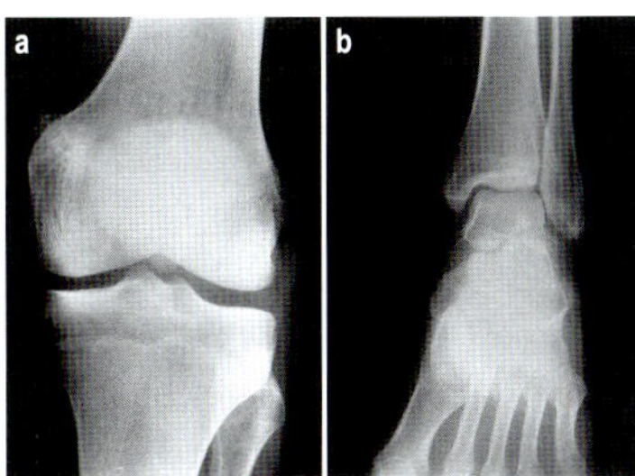

Radiographs of the left (a) superior and (b) inferior tibiofibular joints

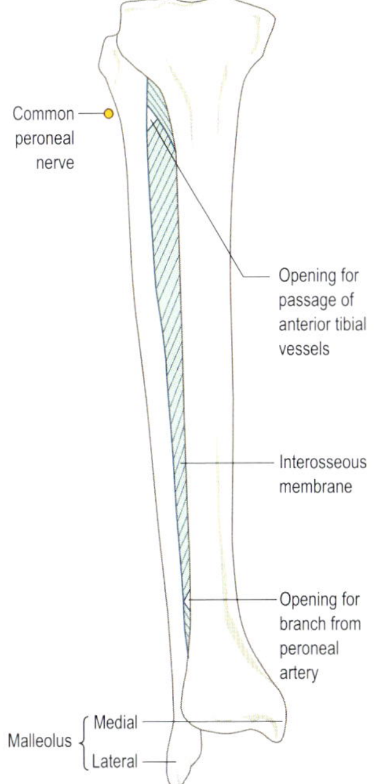

Right tibia, fibula and interosseous membrane

There is no active movement between the tibia and fibula, passive movement being mechanically linked with that at the ankle joint. The tibia articulates proximally at the knee joint: both bones contribute to the ankle joint. The bones articulate by a plane synovial joint superiorly and a fibrous joint (syndesmosis) inferiorly.

Tibia Large medial weight-bearing bone of calf (leg) expanded proximally as the medial and lateral condyles. Below these are the tibial tuberosity anteriorly and medial malleolus distally. The triangular shaft has three borders and three surfaces, the medial surface being superficial throughout its length. On the upper part of the posterior surface is the soleal line. The distal surface is smooth and continuous medially with the malleolar articular surface.

Fibula Long slender lateral bone of calf (leg) surrounded by muscles. It has a proximal expanded head and distal flattened lateral malleolus with a deep malleolar fossa posteromedially.

Palpation

Tibia: The tibial tuberosity can be palpated at the upper end of the anterior border and the condyles 2 cm higher. The medial surface is palpable as far as the medial malleolus and the anterior border throughout its length. The medial malleolus is easily palpated on its medial surface, borders and tip.

Fibula: The head can be palpated posterolaterally below the knee. At the ankle the lateral malleolus can be palpated easily.

ARTICULAR SURFACES

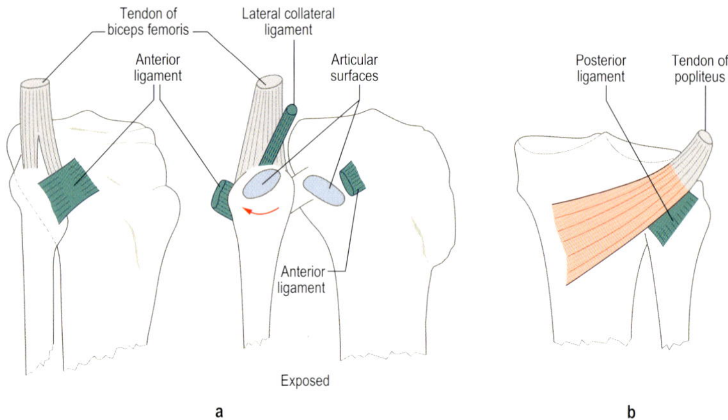

Right superior tibiofibular joint: (a) anterolateral and (b) posterior views

Superior tibiofibular joint Plane synovial joint between an oval facet on the head of fibula and a similar facet on the posterolateral aspect of the undersurface of the lateral tibial condyle.

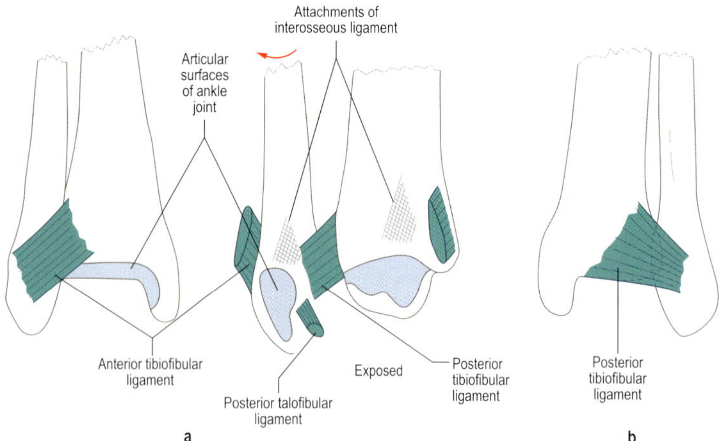

Right inferior tibiofibular joint: (a) anterior and (b) posterior views

Inferior tibiofibular joint Fibrous joint (syndesmosis) between a rough triangular convex surface on the medial aspect of the lower end of fibula and a corresponding area (fibular notch) on the lateral side of tibia.

LIGAMENTS

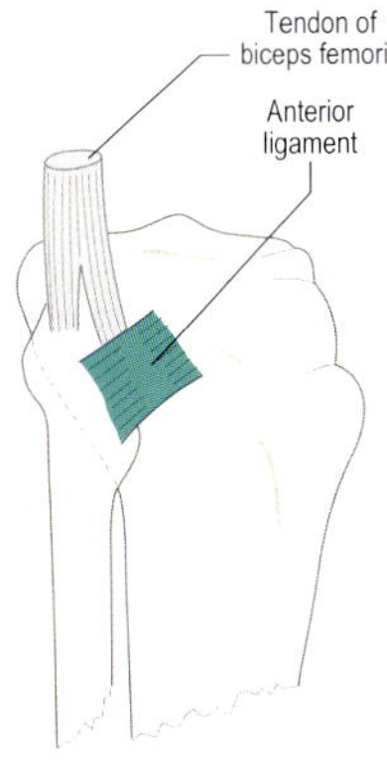

Superior tibiofibular joint Short thick fibrous bands form the *anterior ligament of head of fibula* passing obliquely upwards and medially between the fronts of the fibular head and lateral tibial condyle.

The *posterior ligament of head of fibula* is a single fibrous band passing in a similar direction between the fibular head and back of the lateral tibial condyle.

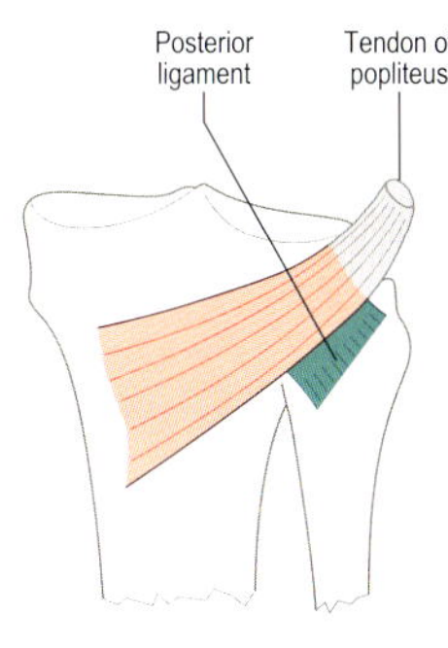

Right superior tibiofibular joint: (a) anterolateral and (b) posterior views

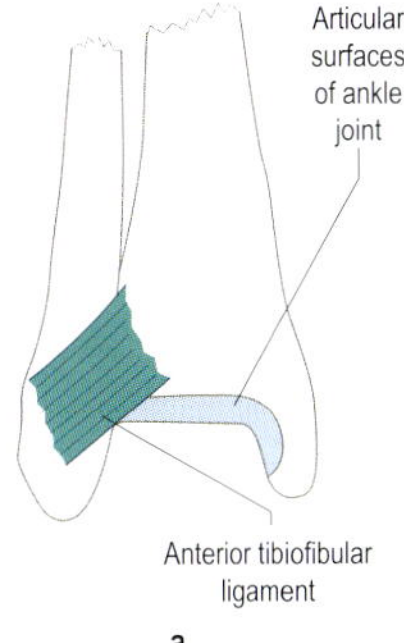

Inferior tibiofibular joint A strong *interosseous ligament*, with short fibrous bands passing inferolaterally, unites the two bones.

Anterior and *posterior tibiofibular ligaments* pass from the borders of the fibular notch to the anterior and posterior surfaces of the lateral malleolus. A *transverse tibiofibular ligament*, deep to the posterior ligament, attaches the length of the posteroinferior tibial surface and upper part of the malleolar fossa.

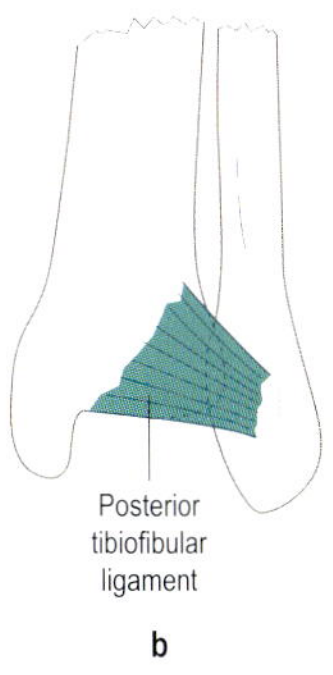

Right inferior tibiofibular joint: (a) anterior and (b) posterior views

Interosseous membrane This runs between the interosseous borders of tibia and fibula, with fibres passing inferolaterally. It does not reach the superior tibiofibular joint but is continuous with the interosseous ligament of the inferior tibiofibular joint. A superior opening transmits the anterior tibial vessels and an inferior opening the peroneal artery. It separates and gives attachment to muscles of the anterior and posterior compartments of the leg.

MOVEMENTS

Automatic movements at the superior and inferior tibiofibular joints accompany dorsiflexion and plantarflexion at the ankle joint: although small they are important. These movements are initiated at the inferior tibiofibular joint and are transmitted to the superior tibiofibular joint.

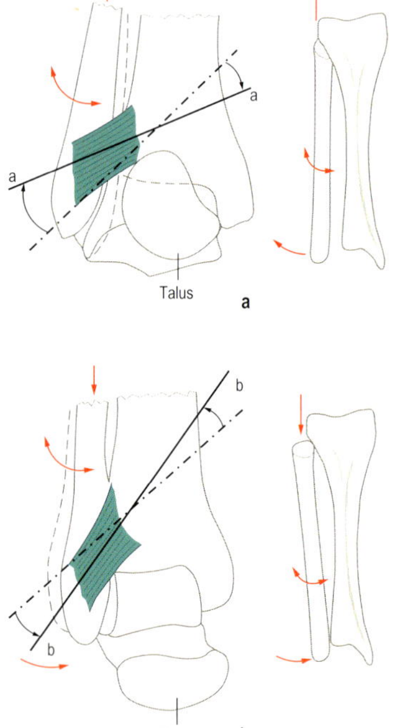

During *ankle dorsiflexion* the malleoli become forced apart due to the broader anterior part of the talar trochlear surface moving into the narrower posterior part of the tibiofibular socket, increasing tension in the anterior and posterior tibiofibular and interosseous ligaments. Because fibres in these ligaments run inferolaterally the fibula tends to be lifted superiorly. There are no bony constraints to this movement as the lateral surface of talus is concave superoinferiorly and convex anteroposteriorly, the latter imparting a slight medial rotation to the tibia.

During *ankle plantarflexion* the malleoli become approximated, partly due to tension developed in the ligaments but also due to the action of tibialis posterior, especially in full plantarflexion. As the lateral malleolus moves medially it also moves inferiorly and rotates in the opposite direction.

Movements (red arrows) of fibula during (a) dorsiflexion and (b) plantarflexion of the ankle joint. – · – · – direction of fibres in tibiofibular and interosseous ligaments in the neutral position; – – direction of same ligament fibres in (*aa*) full dorsiflexion and (*bb*) full plantarflexion

ANKLE

BONES

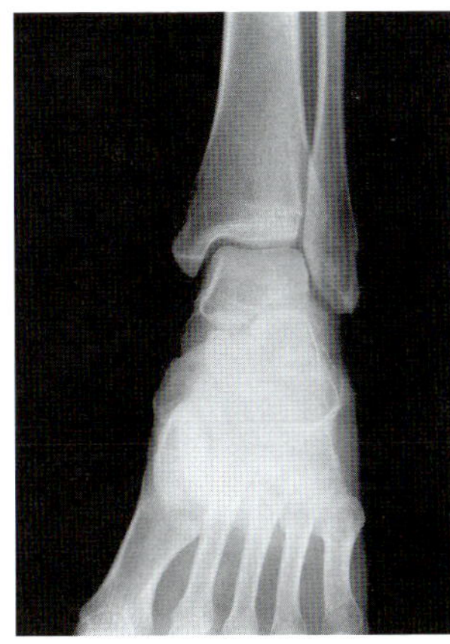

Radiograph of left ankle joint

The ankle is a synovial hinge joint between the distal ends of tibia and fibula and the superior (trochlear) surface of talus.

The axis of the ankle joint is 20–25° to the frontal plane, passing posteriorly as it runs from medial to lateral. Simultaneous movements at both knee and ankle can only be achieved in conjunction with other joints.

Movement of the foot at the ankle joint is rarely performed alone; it is usually combined with subtalar and midtarsal joint motion.

Forward and backward fluctuation of the line of gravity, which usually falls in front of the ankle, is regulated by muscle action to keep it within the base of support.

Palpation

The medial and lateral malleoli can easily be palpated, the lateral being larger and extending further distally.

The *ankle joint line* runs horizontally 1 cm above the tip of the medial malleolus and 2 cm above the tip of the lateral malleolus. It can be palpated on its dorsal surface where the distal end of tibia can be felt between the extensor tendons.

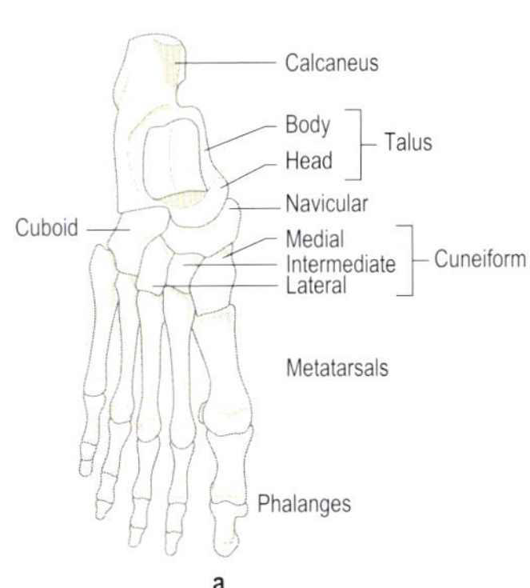

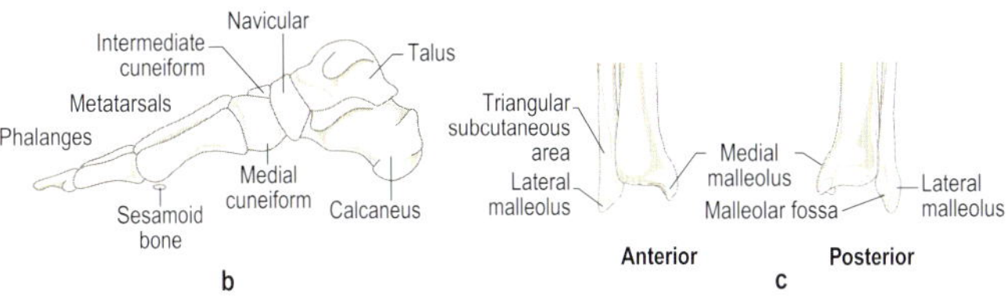

Right foot: (a) superior and (b) medial views; (c) right distal tibia and fibula

ARTICULAR SURFACES

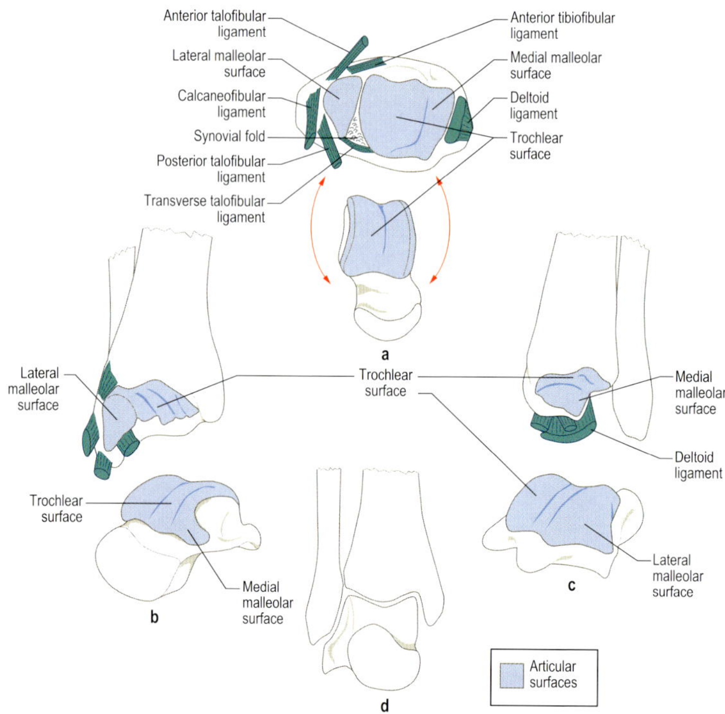

Articular surfaces of ankle joint: (a) trochlear surfaces of tibia and talus; (b) medial and (c) lateral anterior oblique views; (d) coronal section

Tibia The articular surface on the distal end of tibia is continuous with that on the lateral surface of the medial malleolus. The trochlear surface is concave anteroposteriorly, wider anteriorly, and slightly convex transversely with a blunt sagittal ridge either side of which are medial and lateral gutters. The posterior part of the surface projects downwards.

Fibula Triangular articular surface on the medial side of the lateral malleolus with inferior convex apex.

Talus The trochlear surface is convex anteroposteriorly with a central groove, bounded by medial and lateral lips, broader in front than behind: slightly concave transversely. The comma-shaped (tail posterior) medial surface is nearly plane except anteriorly where it inclines medially. The larger triangular (apex inferior) lateral surface runs obliquely anterolaterally, being concave superoinferiorly and anteroposteriorly.

CAPSULE, SYNOVIAL MEMBRANE AND LIGAMENTS

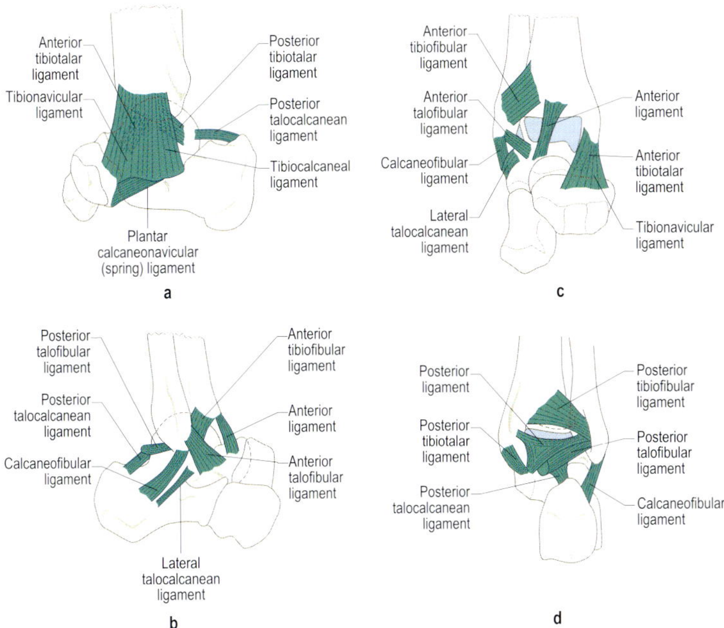

(a) Deltoid and (b) lateral collateral ligaments of ankle joint; (c) anterior and (d) posterior capsular thickenings

A fibrous capsule, strengthened anteriorly and posteriorly, completely surrounds the joint. Synovial membrane lines the capsule and extends superiorly between tibia and fibula as far as the interosseous ligament of the inferior tibiofibular joint.

Deltoid ligament Strong triangular ligament, the apex attaching to anterior and posterior borders and tip of the medial malleolus. It is divided into two deep and two superficial parts. The deeper *anterior* and *posterior tibiotalar ligaments* attach to the medial side of neck and body of talus, blending with the joint capsule. The superficial *tibionavicular* and *tibiocalcaneal ligaments* have a continuous attachment from the navicular tuberosity to the sustentaculum tali, including the 'spring' ligament.

Lateral collateral ligament Comprises three parts: *anterior talofibular ligament* runs between anterior border of lateral malleolus and talar neck; *posterior talofibular ligament* runs horizontally from malleolar fossa to lateral posterior tubercle of talus, both blending with the joint capsule. The cord-like *calcaneofibular ligament* passes from the tip of lateral malleolus to the lateral calcaneal surface behind the peroneal tubercle and blends with the joint capsule.

MOVEMENTS

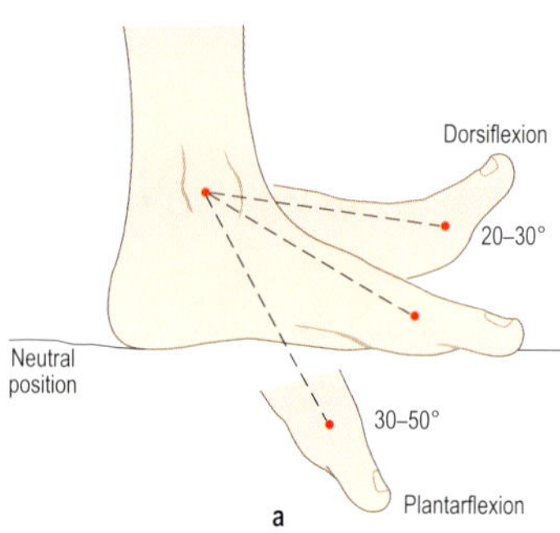

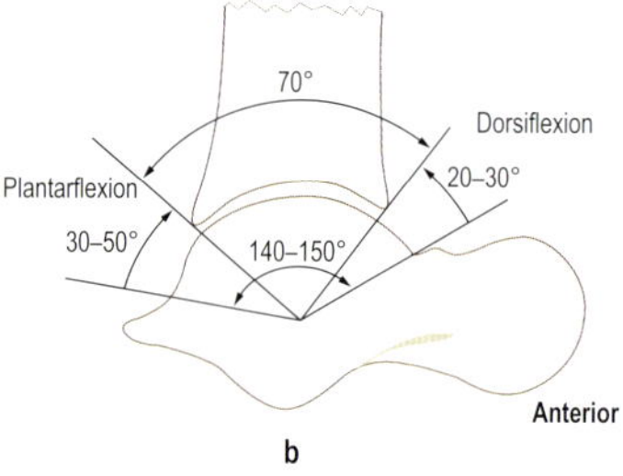

(a) Plantarflexion and dorsiflexion of ankle joint; (b) range of movement determined by articular profiles

Movement occurs about a transverse axis level with the tip of the lateral malleolus: the axis changes slightly during movement because the trochlear surface of talus is elliptical.

In the neutral position the foot makes a right angle with the leg. In *dorsiflexion* the foot is drawn towards the leg: *plantarflexion* is movement in the opposite direction. The range of dorsiflexion (20–30°) and plantarflexion (30–60°) is determined by the profiles of the articular surfaces and shows considerable variation between individuals.

In *dorsiflexion* the broader anterior part of the trochlear surface of talus moves into the narrower tibiofibular mortise, causing separation of the malleoli and increasing tension in ligaments of the inferior tibiofibular joint. In this position the ankle joint is most stable.

In *plantarflexion* the narrower posterior part of the talus moves forwards into the broader tibiofibular mortise: this is the least stable position of the joint.

Accessory movements: longitudinal distraction and anterior and posterior gliding of the talus in the mortise.

FOOT

INTRODUCTION

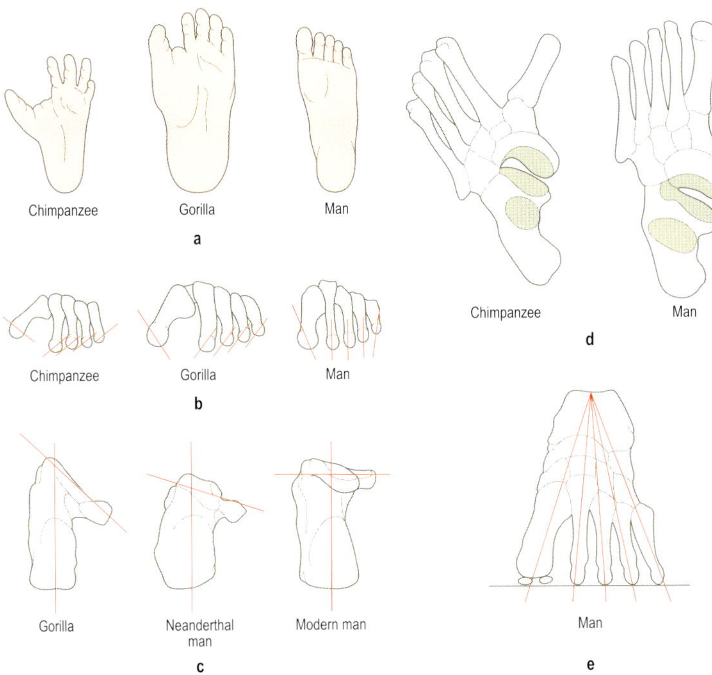

(a) Transition to human foot; (b) loss of rotation of metatarsals during evolution of the foot; (c) changes in obliquity of the sustentaculum tali; (d) reduced angulation of the great toe; (e) anterior support of the human foot

The human foot has evolved from the mobile, prehensile organ of many primates to the specialized supporting structure necessary for bipedal locomotion and has lost the ability to oppose the great toe. The metatarsal heads are no longer rotated towards each other as required for gripping but directed anteroposteriorly, with the axis of leverage shifting medially to lie between the 1st and 2nd metatarsals. As a result the medial border of the foot has become flattened and depressed and a transverse arch developed so that all metatarsal heads now support the forefoot. The axis of abduction/adduction is through the 2nd digit compared to the 3rd in the hand.

With the adoption of bipedalism the calcaneus has become more massive, especially the sustentaculum tali which has assumed a more horizontal orientation in order to support the talus and superincumbent body weight.

BONES

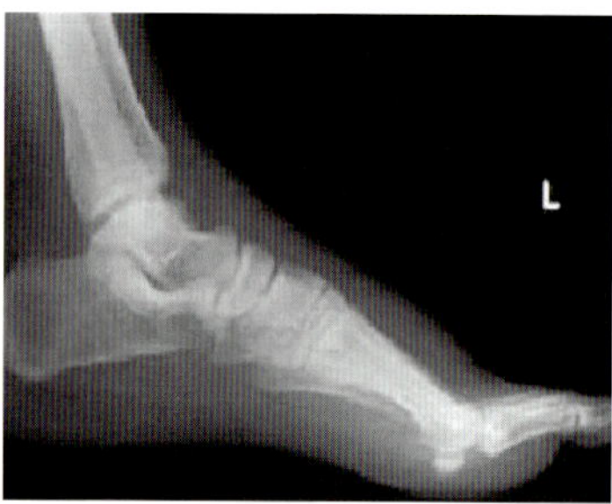

Radiograph of the foot

The foot consists of several small bones: the *tarsus* posteriorly and the *metatarsals* and *phalanges* anteriorly. The tarsus and metatarsals comprise the foot proper and the phalanges the toes. The tarsal bones are: *calcaneus* (heel), *talus*, *navicular*, three *cuneiforms* and *cuboid*.

Palpation

The calcaneus is subcutaneous on its lateral, posterior and medial aspects: 1 cm below the medial malleolus is the sustentaculum tali, in front of which is the tuberosity of navicular. The head and neck of talus can be gripped in the two hollows anteroinferior to the medial malleolus. Midway along the lateral border of the foot the base of the 5th metatarsal can be felt. The metatarsal shafts and heads can be palpated on the dorsum of the foot.

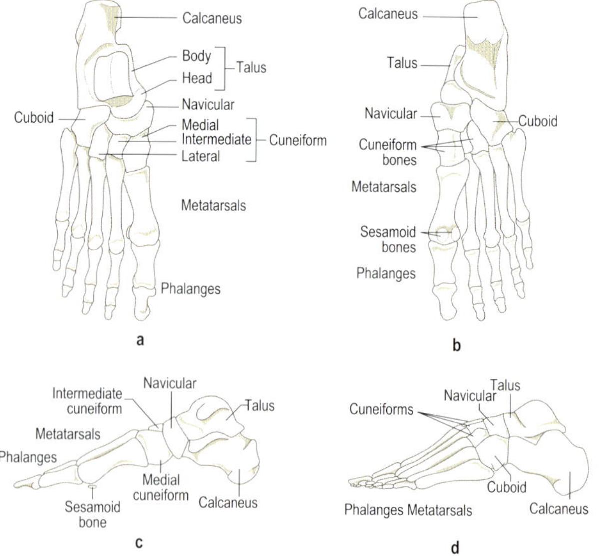

Right foot: (a) superior, (b) inferior and (c) medial views; (d) left foot lateral view

JOINTS

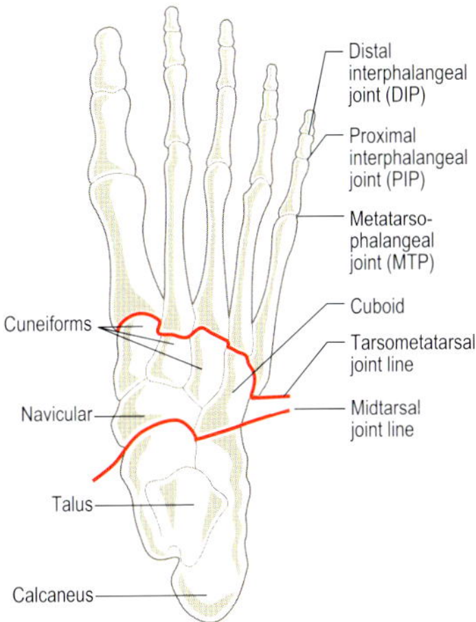

The midtarsal and tarsometatarsal joint lines

Subtalar joint Synovial articulation between the inferior surface of the body of talus and the superior surface of calcaneus.

Midtarsal joint Combination of the separate medial talocalcaneonavicular and the lateral calcaneocuboid joints forming an irregular articulation across the foot, with the calcaneus and talus behind and navicular and cuboid in front. The talocalcaneonavicular joint is the articulation of the head of talus with the posterior surface of the navicular and deep surface of the plantar calcaneonavicular ('spring') ligament: the calcaneocuboid joint is between the anterior surface of the calcaneus and the posterior surface of the cuboid.

Tarsometatarsal joint Articulation between the cuboid and three cuneiforms posteriorly and the bases of all five metatarsals anteriorly. The joint line is irregular and arched, with the medial end being 2 cm anterior to the lateral end.

Joints also exist between the navicular and cuneiform bones (*cuneonavicular*), as well as between the cuneiforms (*intercuneiform*), the lateral cuneiform and cuboid (*cuneocuboid*), metatarsal bases (*intermetatarsal*), the metatarsal heads and proximal phalanges (*metatarsophalangeal, MTP*) and the phalanges (*proximal interphalangeal, PIP*, and *distal interphalangeal, DIP*).

ARTICULAR SURFACES OF TARSAL JOINTS

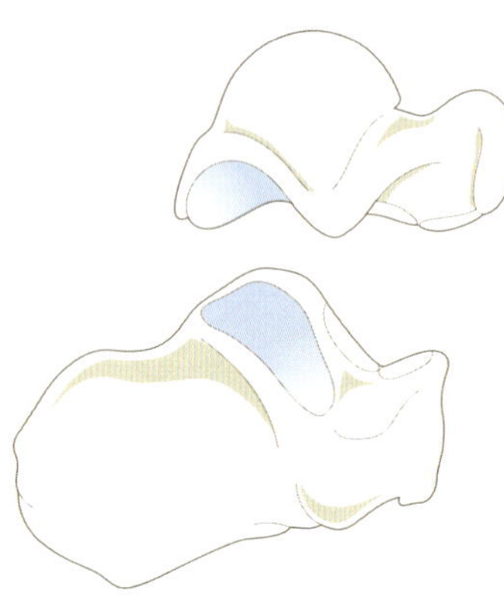

Subtalar joint surfaces

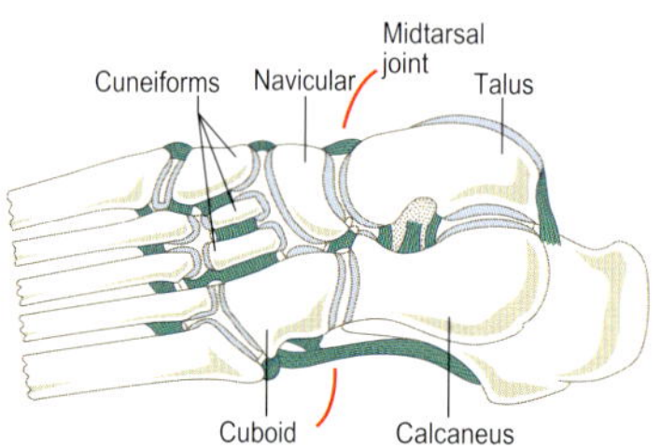

Mid (transverse) tarsal and tarsometatarsal joints

Subtalar joint Plane synovial joint between an oval articular facet on the superior surface of the calcaneus and a similar facet on the undersurface of the body of talus. The facet on the calcaneus is concavoconvex about its long axis, matching the reciprocally shaped facet on the talus.

A thin loose fibrous capsule surrounds the joint reinforced by the talocalcanean ligaments.

Midtarsal joint Medially a synovial ball-and-socket joint between the head and lower surface of the neck of talus and navicular anteriorly and the plantar calcaneonavicular ligament inferiorly; laterally a plane synovial joint between the square facets on the calcaneus and cuboid which have reciprocally shaped concavoconvex surfaces. A fibrous capsule, lined by synovial membrane, surrounds the joint thickened above and below by dorsal and plantar calcaneocuboid ligaments.

Tarsometatarsal joints Plane synovial joints which overlap each other. The base of the 1st metatarsal articulates with the medial cuneiform in its own joint cavity separate from the other tarsometatarsal joints. The base of the 2nd is held in a mortise formed by all three cuneiforms and articulates with them. The base of the 3rd articulates with the lateral cuneiform, the 4th with the cuboid and a small part of the lateral cuneiform, and the 5th with the cuboid.

The joint capsule is formed by thickenings of the interosseous, plantar and dorsal tarsometatarsal ligaments. Synovial membrane attaches to the articular margins and lines all non-articular surfaces.

LIGAMENTS OF TARSAL JOINTS

Subtalar joint

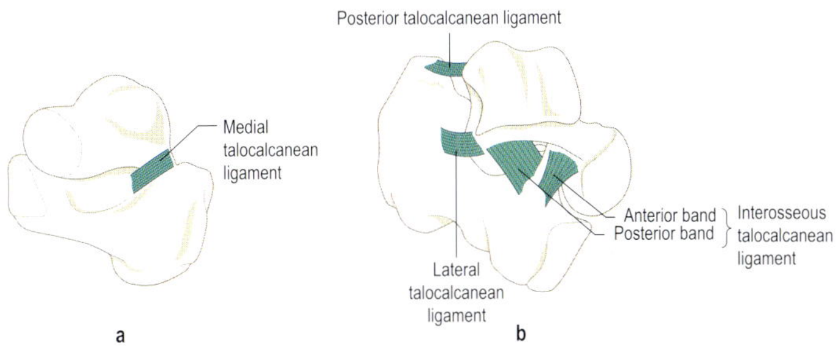

Talocalcanean ligaments: (a) medial and (b) lateral aspects

Interosseous ligament Two thick bands attaching to the floor of the sinus tarsi: anterior runs obliquely superiorly, anteriorly and medially to the neck of talus; posterior runs superiorly, posteriorly and laterally to talus anterior to the joint.
Medial talocalcanean ligament From the medial posterior talar tubercle to the posterior border of the sustentaculum tali.
Posterior talocalcanean ligament From the lateral talar tubercle to the superomedial calcaneal surface.
Lateral talocalcanean ligament From the lateral talar tubercle to the lateral surface of calcaneus parallel and deep to the calcaneofibular ligament.

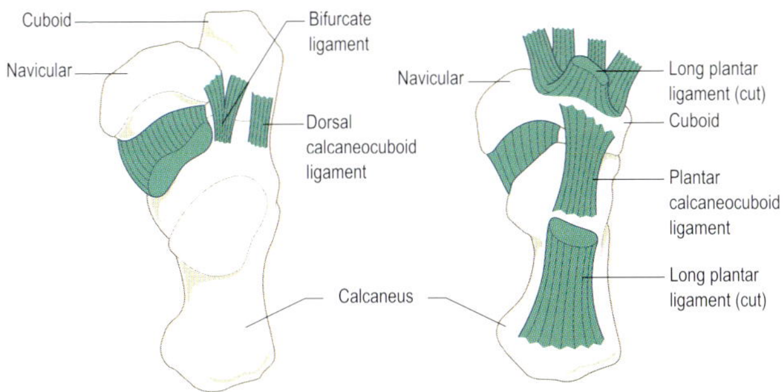

Calcaneocuboid ligaments

Calcaneocuboid joint

Plantar calcaneocuboid (short plantar) ligament Strong short broad band from rounded eminence on anteroinferior surface of calcaneus to plantar surface of cuboid behind ridge of peroneal groove: reinforces joint capsule.
Dorsal calcaneocuboid ligament Thin broad band strengthening dorsal aspect of capsule.

LIGAMENTS OF TARSAL JOINTS

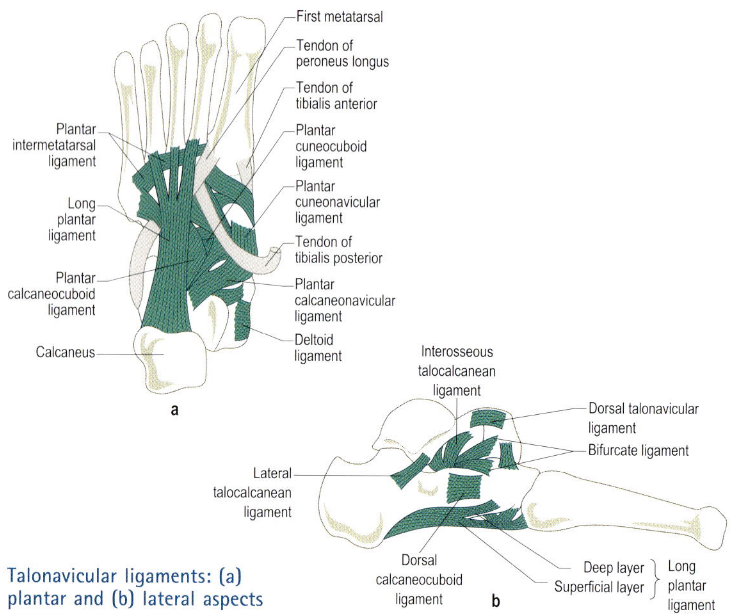

Talonavicular ligaments: (a) plantar and (b) lateral aspects

Talonavicular joint

Plantar calcaneonavicular ('spring') ligament Thick, dense, strong fibroelastic ligament from the anterior end and medial border of the sustentaculum tali to the medial end and tuberosity of navicular. Blends with and is supported by the deltoid ligament: upper surface is fibrocartilaginous for articulation with the talar head.
Dorsal talonavicular ligament From the talar neck to the dorsal surface of navicular: reinforces joint capsule.

Ligaments of midtarsal joint

Bifurcate ligament From the upper calcaneal surface anterior to the sinus tarsi: calcaneonavicular part runs upwards to lateral surface of navicular; calcaneocuboid part runs horizontally to the dorsomedial angle of cuboid; strong connection between the first and second row of tarsal bones.
Long plantar ligament Attached posteriorly between anterior and posterior calcaneal tubercles: deeper fibres attach to the ridge on cuboid, intermediate fibres to cuboid tuberosity and superficial fibres attach to lateral four metatarsal bases.
Dorsal, plantar, interosseous cuboideonavicular ligaments Also pass between the cuboid and navicular.

LIGAMENTS OF TARSOMETATARSAL AND MTP JOINTS

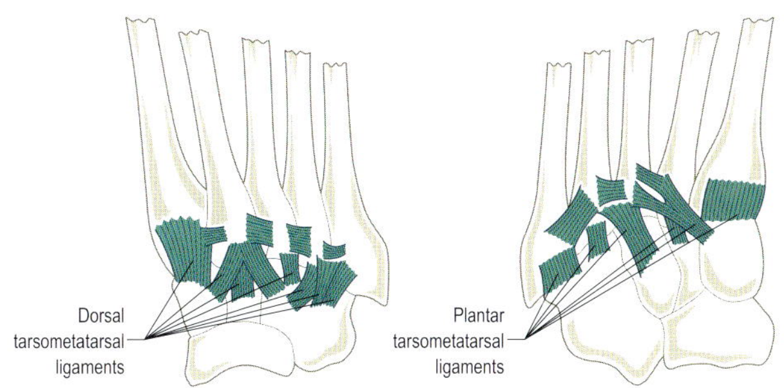

Dorsal and plantar tarsometatarsal ligaments

Tarsometatarsal joints

Dorsal tarsometatarsal ligaments Weak short slips between adjacent dorsal surfaces of tarsal bones and metatarsals.

Plantar tarsometatarsal ligaments Similar slips connecting plantar surfaces.

Interosseous tarsometatarsal ligaments First ligament runs between anterolateral surface of medial cuneiform and base of 2nd metatarsal, second between anterolateral angle of lateral cuneiform and base of 4th metatarsal. Occasionally third ligament runs between lateral side 2nd metatarsal and lateral cuneiform.

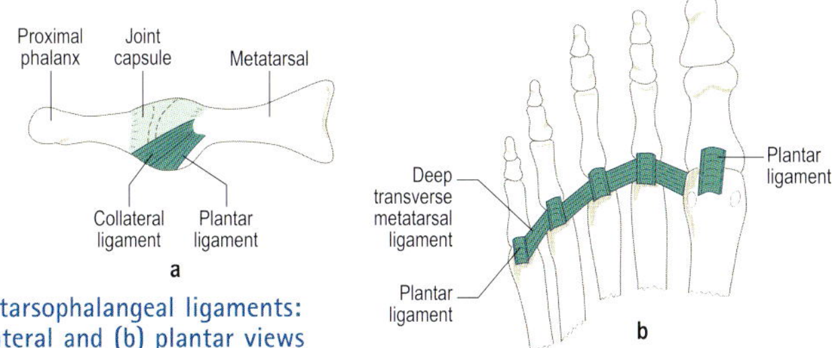

Metatarsophalangeal ligaments: (a) lateral and (b) plantar views

Metatarsophalangeal (MTP) joints

Collateral ligaments From each side of metatarsal head to phalangeal base.

Plantar ligament Dense fibrocartilaginous plate from base of proximal phalanx to sides of collateral and deep transverse metatarsal ligaments: in great toe incorporates sesamoid bones.

Deep transverse metatarsal ligament Connects metatarsal heads, joint capsules and plantar ligaments of all metatarsals.

MOVEMENTS

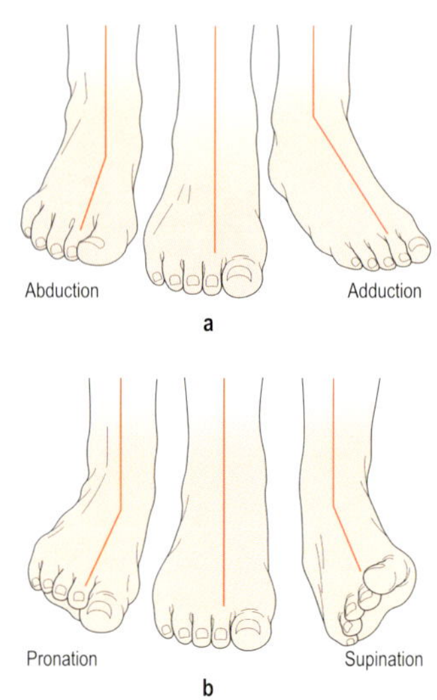

(a) Abduction and adduction about long axis of leg; (b) pronation and supination about longitudinal axis of foot

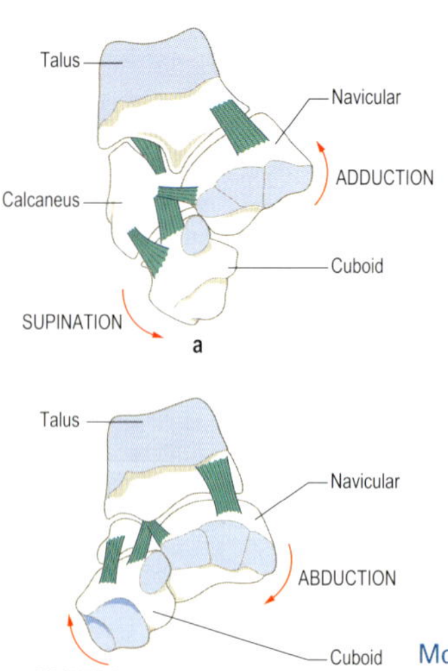

Movements of talus, calcaneus, cuboid and navicular in (a) inversion and (b) eversion

Combined movements of the foot at the subtalar and midtarsal joints produce inversion and eversion. Because the talus participates in both joints, adduction (inversion) at the subtalar joint is always accompanied by supination at the midtarsal joint: similarly abduction (eversion) is always accompanied by pronation.

In *inversion* the foot is twisted, raising the medial and lowering the lateral border to turn the sole to face medially: ankle plantarflexion increases the range of movement. *Eversion* is the opposite so that the lateral border is raised and medial lowered turning the sole laterally: ankle dorsiflexion increases the range of the movement.

During inversion and eversion the combined movements of the calcaneus, cuboid and navicular, with respect to a fixed talus, occur about a fixed axis passing obliquely upwards, forwards and medially, passing through the sinus tarsi.

In inversion the navicular and cuboid are pulled medially so that the forefoot moves anteriorly and medially. At the same time they also rotate about an anteroposterior axis through the bifurcate ligament, raising the navicular and lowering the cuboid. Elevation of the medial and depression of the lateral longitudinal arches turn the sole to face medially. In eversion the navicular and cuboid are pulled laterally so that the forefoot moves anteriorly and laterally. At the same time they rotate about the same axis, raising the cuboid and lowering the navicular, causing the sole to face laterally.

ANKLE AND FOOT

MUSCLES

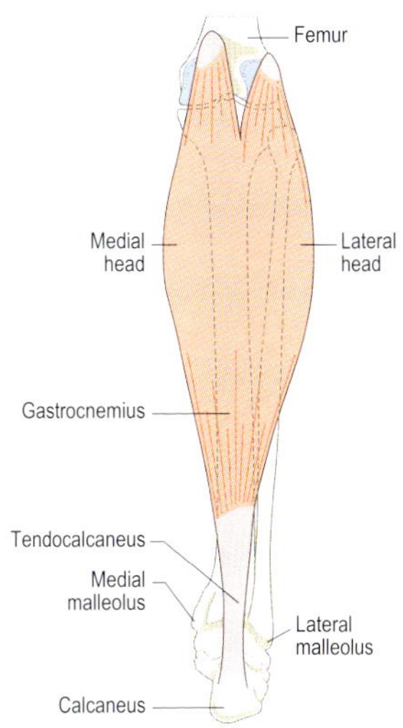

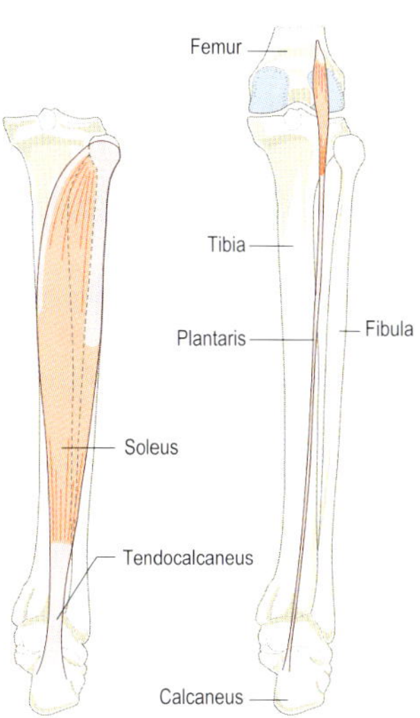

Posterior views right calf

Gastrocnemius – most superficial calf muscle: plantarflexes ankle and flexes knee.

Origin: medial head – medial supracondylar ridge and adductor tubercle of femur; lateral head – outer surface lateral condyle of femur.

Insertion: via tendocalcaneus to posterior surface calcaneus.

Nerve supply: tibial nerve S1, 2.

Soleus – deep to gastrocnemius: plantarflexes ankle.

Origin: soleal line posterior surface of tibia; posterior surface upper third fibula and fibrous arch between.

Insertion: via tendocalcaneus with gastrocnemius into posterior surface calcaneus.

Nerve supply: tibial nerve S1, 2.

Plantaris – slender muscle running from lateral supracondylar ridge of femur to join tendocalcaneus.

Nerve supply: tibial nerve S1, 2.

Tendocalcaneus (Achilles tendon) is thickest, strongest tendon in body and is subjected to great force during powerful movements lifting body weight, e.g. jumping. During such movements it is liable to rupture. It is enclosed in a 'paratenon' which may become inflamed – Achilles tendonitis.

Gastrocnemius acts as a powerful propelling muscle whereas soleus has a more postural antigravity role, e.g. preventing body swaying forwards during standing. Gastrocnemius is able to flex knee via its attachments to femur.

Plantaris contributes little to movement but its muscle belly is prone to damage.

The two bellies of gastrocnemius are easily palpated in upper third of calf with soleus palpable in lower half.

Historically, gastrocnemius and soleus were called triceps surae because they share the single tendocalcaneus.

MUSCLES

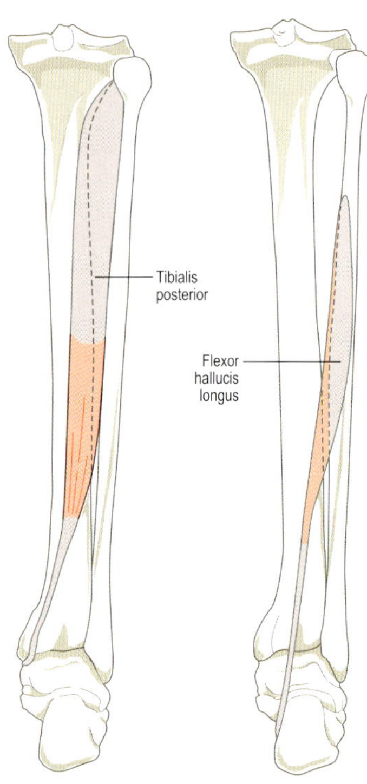

Posterior view right leg

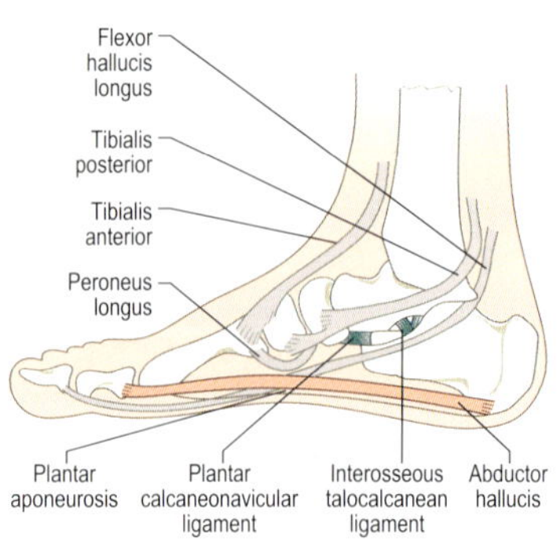

Medial view right foot

Tibialis posterior – inverts and plantarflexes foot; deepest muscle in the calf.
Origin: upper part posterior surface of tibia (below soleal line), posterior surface of fibula and adjacent interosseous membrane.
Insertion: tubercle of navicular and medial cuneiform with expansions to all tarsal bones except talus.
Nerve supply: tibial nerve L4, 5.

Flexor hallucis longus – plantarflexes big toe and then ankle.
Origin: unipennate muscle arising lower two-thirds posterior surface of fibula.
Insertion: plantar surface base of distal phalanx.
Nerve supply: tibial nerve S1, 2.

The tendon of tibialis posterior passes behind the medial malleolus and deep to the flexor retinaculum enclosed in its own synovial sheath. That of flexor hallucis longus also passes below the flexor retinaculum within its own synovial sheath and runs in grooves on the tibia, back of the talus and sustentaculum tali.

Tibialis posterior has an important role in maintaining balance and sustaining the arches of the foot.

Flexor hallucis longus is a strong muscle and produces the final thrust from the foot during the 'push-off' phase of walking. It also helps maintain the medial longitudinal arch.

MUSCLES

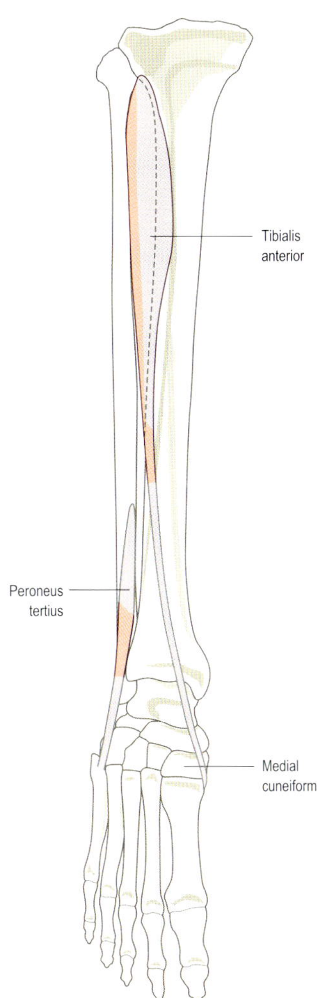

Anterior view right leg

Tibialis anterior – dorsiflexes and inverts the foot.

Origin: upper two-thirds lateral surface of tibia and adjacent interosseous membrane.

Insertion: medial side of medial cuneiform and base 1st metatarsal.

Nerve supply: deep peroneal nerve L4, 5.

Tibialis anterior is a fusiform muscle whose tendon runs below the superior and inferior extensor retinacula, enclosed in its own synovial sheath.

Palpation The tendon is easily seen and felt on the anteromedial part of the ankle.

Functionally it acts as a stabilizer, balancing the body and maintaining the medial longitudinal arch of foot. During the swing phase of gait it acts to prevent the toes catching the ground and on heel strike lowers the foot (eccentrically) to the floor. If paralysed, the characteristic 'foot drop' gait results.

Tibialis anterior is one of the four 'anterior tibial muscles' contained in a tight osseofascial compartment. Swelling or hypertrophy in this limited space through overuse is a common cause of 'compartment syndrome'.

Peroneus (fibularis) tertius – dorsiflexes and everts foot.

Origin: lower quarter anterior surface fibula.

Insertion: dorsal aspect base 5th metatarsal.

Nerve supply: deep peroneal nerve L5, S1.

The tendon passes below superior and inferior extensor retinacula with extensor digitorum longus, enclosed in a synovial sheath.

The position of the tendon is close to the anterior talofibular ligament of ankle and it is possible that this muscle helps prevent injury during excessive inversion.

MUSCLES

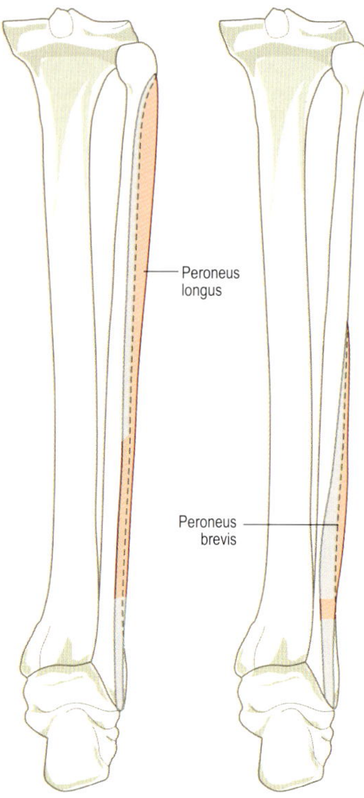

Posterior view right leg

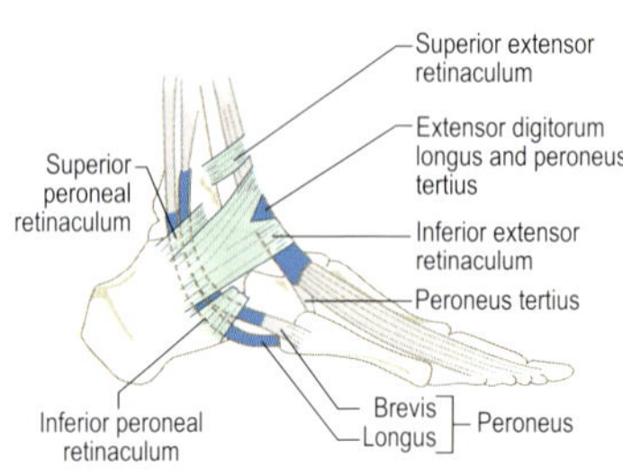

Lateral view right foot

Peroneus (fibularis) longus – everts and plantarflexes the foot.
Origin: upper two-thirds lateral surface and head of fibula.
Insertion: plantar surface medial cuneiform and adjacent base 1st metatarsal.
Nerve supply: superficial peroneal nerve L5, S1.

Peroneus (fibularis) brevis – everts and plantarflexes the foot.
Origin: lower two-thirds lateral surface fibula.
Insertion: tubercle on lateral base 5th metatarsal.
Nerve supply: superficial peroneal nerve L5, S1.

Tendons of both muscles are enclosed in synovial sheaths and pass in a groove behind the lateral malleolus with brevis being next to the bone. They are held in position here by the superior peroneal retinaculum before crossing the calcaneus where they pass above (brevis) and below (longus) the peroneal tubercle. Longus then turns medially to cross the foot, grooving the cuboid.

Both muscles help maintain balance by preventing sideways sway.

Brevis supports the lateral ligament of the ankle and can prevent overinversion.

Longus works with tibialis anterior in supporting the medial side of the foot during powerful activities such as running. Occasionally laxity of the retinaculum allows the peroneal tendons to 'snap' out of the groove and lie on the lateral malleolus.

MUSCLES

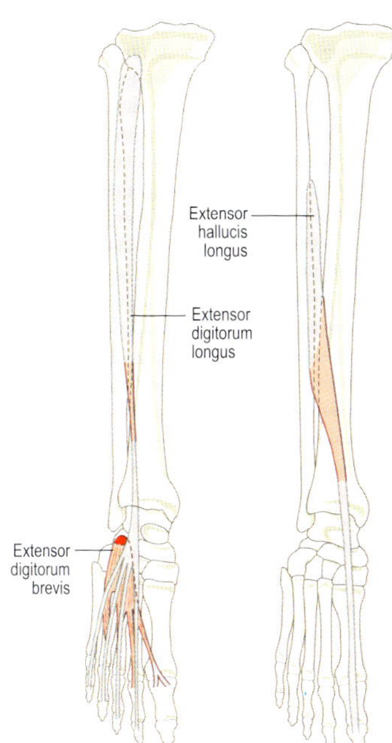

Anterior view right leg

Muscle belly of extensor digitorum brevis can be palpated on the superolateral aspect of foot when toes are dorsiflexed – the only muscle fibres on dorsum of foot.

Extensors digitorum and hallucis longus assist tibialis anterior in holding the foot up during swing phase of gait and lowering it (eccentrically) on heel strike.

Extensor digitorum longus – dorsiflexes in turn, interphalangeal, metatarsophalangeal joints of toes and ankle joint.

Origin: upper two-thirds anterior surface fibula, interosseous membrane and lateral tibial condyle.

Insertion: forms extensor hood (dorsal digital expansion) which attaches to distal and middle phalanx of the lateral four toes on their dorsal surface – the same arrangement as in the hand.

Nerve supply: deep peroneal nerve L5, S1.

The single tendon passes below the superior and inferior retinacula in a synovial sheath and then divides into four.

Lumbrical muscles join the extensor hood on its medial side.

Extensor digitorum brevis – dorsiflexes hallux and adjacent three toes.

Origin: anterior part upper surface calcaneus.

Insertion: lateral side extensor hood 2nd to 4th toes and by tendon into dorsal surface proximal phalanx of hallux (extensor hallucis brevis).

Nerve supply: deep peroneal nerve L5, S1.

Extensor hallucis longus – dorsiflexes joints of hallux and then ankle.

Origin: middle half anterior surface fibula and adjacent interosseous membrane.

Insertion: base dorsal surface distal phalanx of hallux.

Nerve supply: deep peroneal nerve L5, S1.

The tendon passes under superior and inferior extensor retinacula in own synovial sheath.

MUSCLES

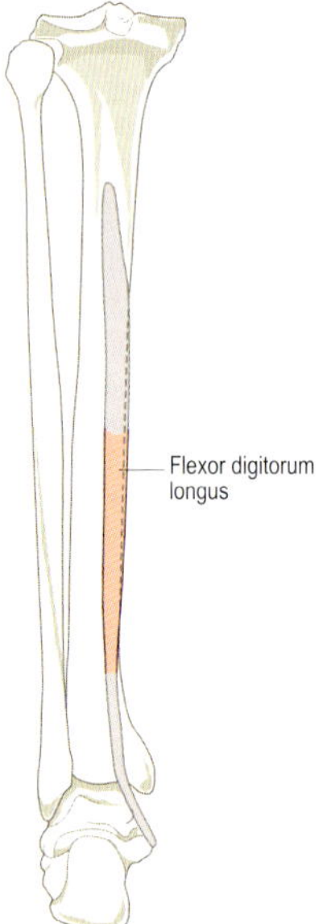

Posterior view left leg

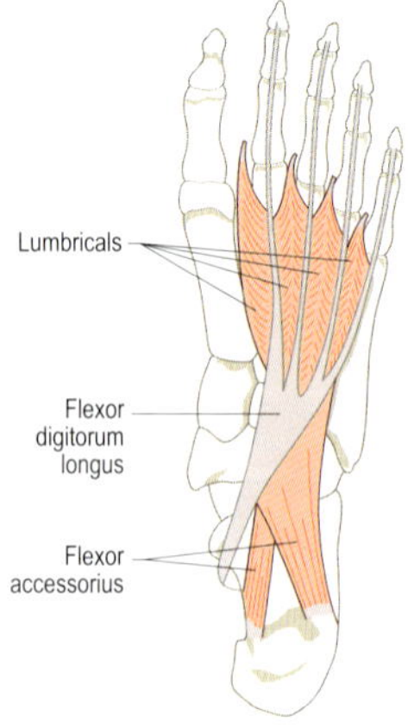

Plantar surface left foot

Flexor digitorum longus – lies deep to soleus and flexes (plantarflexes), in turn, distal and proximal interphalangeal joints (DIP and PIP), metatarsophalangeal (MTP) and ankle joints.
Origin: medial part posterior surface of tibia below soleal line.
Insertion: tendon passes behind medial malleolus and divides into four to insert into plantar surface base distal phalanx of the lateral four toes.
Nerve supply: tibial nerve L5, S1, 2.

> Flexor digitorum longus pierces and passes through flexor digitorum brevis in the same way as superficialis and profundus in the hand.

Flexor accessorius (quadratus plantae) – flexes toes via tendon of flexor digitorum longus (FDL).
Origin: medial and lateral tubercles of calcaneus.
Insertion: tendon of FDL at midpoint of sole.
Nerve supply: lateral plantar nerve S2, 3.

> Functionally this muscle can act on the toes via the tendon of FDL even when the foot is plantarflexed and FDL is already shortened.

Lumbricals – these four muscles flex the MTP joint and extend the IP joints.
Origin: from tendons of FDL: 1st – from medial side of tendon to 2nd toe, the others arise from adjacent tendons of FDL.
Insertion: medial side extensor hood and base proximal phalanx.
Nerve supply: medial lumbrical – medial plantar nerve S1, 2; lateral three lumbricals – lateral plantar nerve S2, 3.

> The main function of the lumbricals is to prevent clawing of the toes during the propulsive phase of gait.

MUSCLES

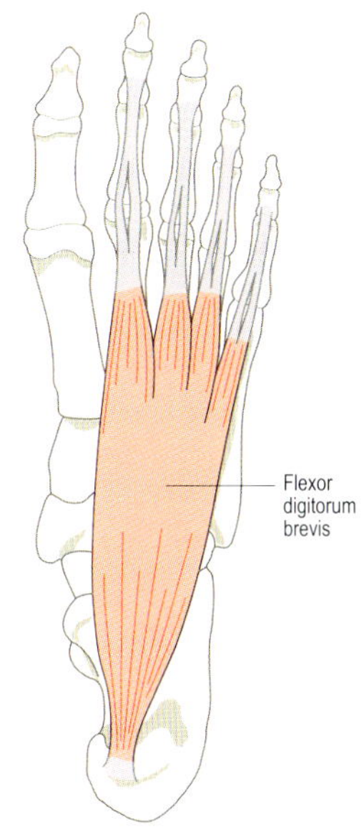

Plantar surface left foot

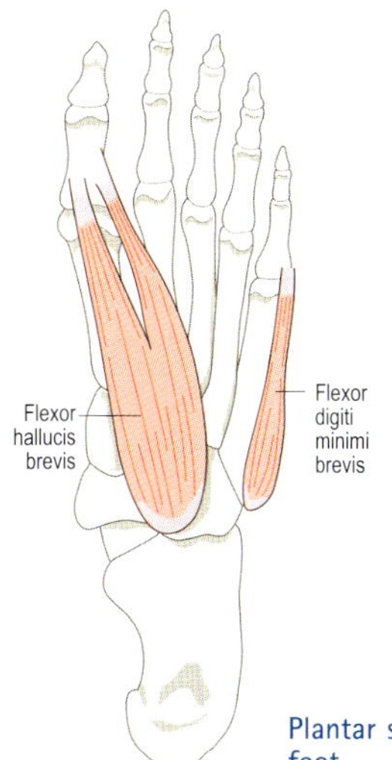

Plantar surface left foot

Flexor digitorum brevis – flexes proximal interphalangeal (PIP) and metatarsophalangeal (MTP) joints of toes.
Origin: medial tubercle calcaneus and plantar aponeurosis.
Insertion: base middle phalanx of lateral four toes.
Nerve supply: medial plantar nerve S2, 3.

The tendons of brevis split at the level of the metatarsophalangeal (MTP) joints to allow the longus tendons to pass through, reuniting deep to longus prior to insertion – the same arrangement as for flexor tendons in the hand.

Flexor hallucis brevis – flexes MTP joint of big toe (hallux).
Origin: medial side plantar surface cuboid and adjacent lateral cuneiform.
Insertion: base proximal phalanx of big toe via two tendons, each containing a sesamoid bone.
Nerve supply: medial plantar nerve S1, 2.

Flexor digiti minimi brevis – flexes MTP joint of little toe.
Origin: plantar aspect base 5th metatarsal.
Insertion: plantar surface proximal phalanx little toe.
Nerve supply: lateral plantar nerve S2, 3.

During gait flexor hallucis brevis assists in the final push-off from the ground, as does flexor digitorum brevis. Both muscles also help support the arches of the foot from below and help cushion its plantar surface.

MUSCLES

a

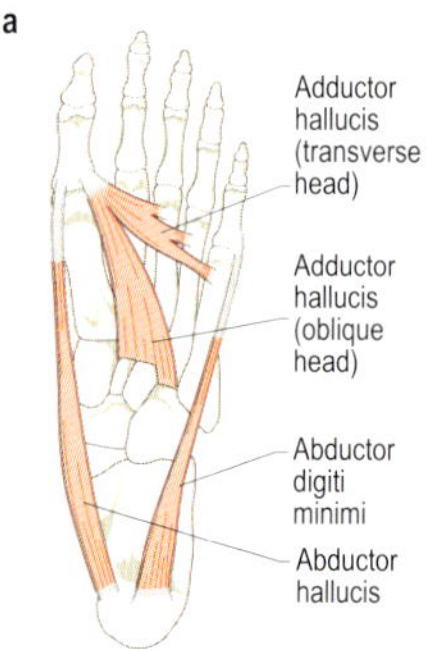

Plantar surface left foot

b

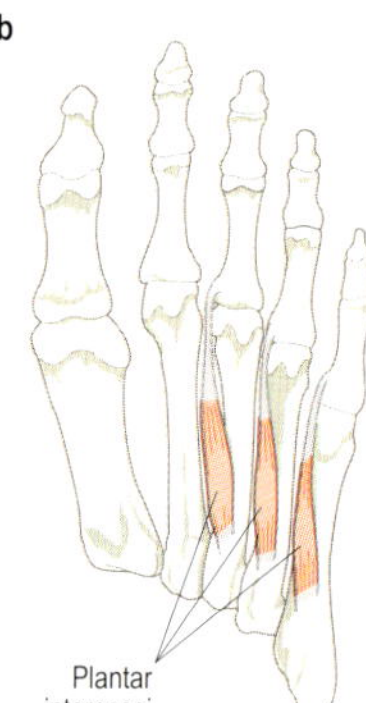

Plantar surface left foot

c

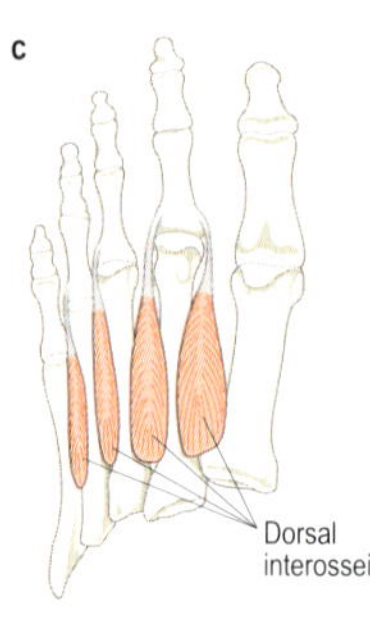

Dorsal surface left foot

Abductor hallucis – abducts hallux (big toe).
Origin: plantar aponeurosis and medial tubercle calcaneus.
Insertion: medial side base proximal phalanx hallux.
Nerve supply: medial plantar nerve S1, 2.

Abductor digiti minimi – abducts little toe.
Origin: tubercles of calcaneus and intermuscular septa.
Insertion: lateral side base proximal phalanx 5th toe.
Nerve supply: lateral plantar nerve S2, 3.

Adductor hallucis – adducts hallux.
Origin: oblique head – plantar surfaces bases 2nd, 3rd, 4th metatarsals; transverse head – plantar surface lateral three metatarsophalangeal joints and their ligaments.
Insertion: lateral side base proximal phalanx hallux.
Nerve supply: lateral plantar nerve S2, 3.

Plantar interossei – these three muscles originate from plantar and medial aspects of 3rd, 4th, 5th metatarsals inserting into medial side base proximal phalanx of same toe. They adduct these three toes, i.e. drawing them towards 2nd toe.

Dorsal interossei – these four bipennate muscles originate from sides of adjacent metatarsals and insert into lateral side proximal phalanx of 2nd, 3rd, 4th toes and medial side 2nd toe. They abduct the toes, i.e. move them away from a line through the 2nd toe. This explains why the 2nd toe has no adductors but two abductors attaching to it.
Nerve supply: all seven interossei – lateral plantar nerve S2, 3.

> Functionally, all these muscles help support the arches of the foot, many by pressing the toes to the ground, supporting the anterior 'pillar'. Adductor hallucis supports the anterior transverse arch.

RELATIONS

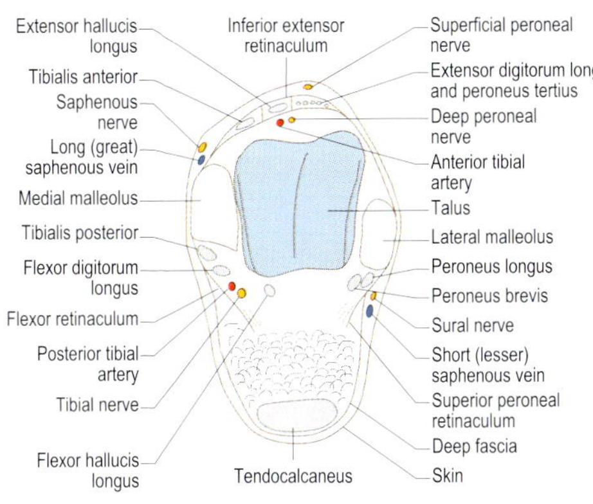

Transverse section through the ankle joint showing related structures

All muscles, vessels and nerves entering the foot cross the ankle joint, mostly deep to retinacula, i.e. extensor, peroneal (superior and inferior) and flexor. Anterior to the extensor retinacula is the superficial peroneal nerve: anterior to the medial malleolus are the long saphenous vein and saphenous nerve. Superficial to the superior peroneal retinaculum are the short saphenous vein and sural nerve.

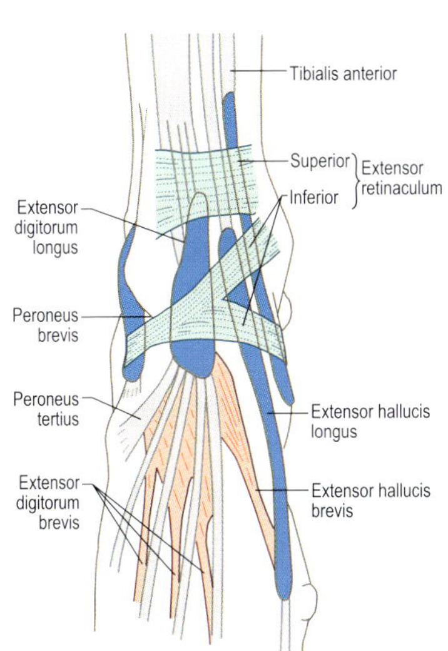

Dorsal view right ankle and foot showing synovial sheaths, retinacula and tendons

Palpation

Tendons of peroneus longus and brevis, tibialis posterior and flexor digitorum longus and all extensor tendons can be palpated crossing the ankle joint.

The dorsalis pedis pulse can be felt between the tendons of extensors hallucis and digitorum longus anteriorly, and the posterior tibial artery pulse behind the medial malleolus posterior to flexor digitorum longus tendon.

ARCHES

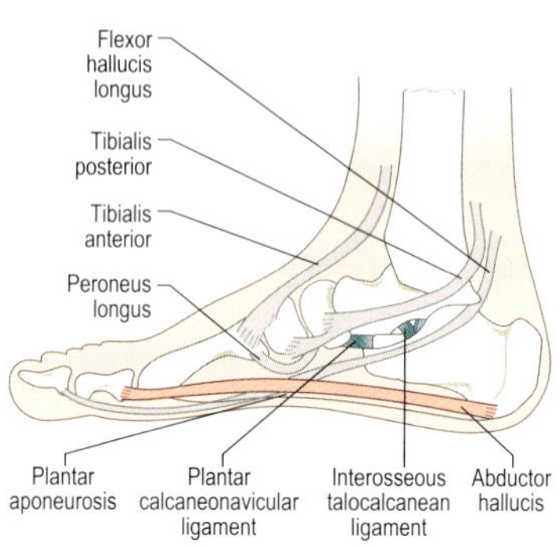

Medial longitudinal arch of foot

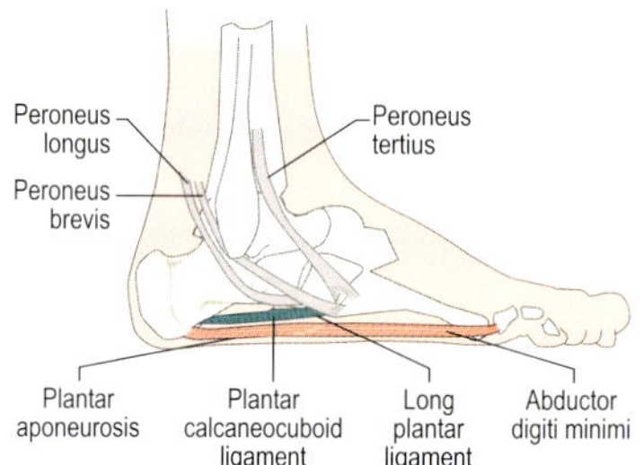

Lateral longitudinal arch of foot

The arched structure of the foot allows the combination of stability and flexibility necessary for function. The *medial longitudinal arch* has the greatest curvature and flexibility whereas the *lateral longitudinal arch* is flatter and more rigid. The bones of these arches are held together strongly by ligaments but require muscle action to sustain their shape. The *plantar aponeurosis* runs from the heel to the toes and supports both longitudinal arches acting as a ligament. Intrinsic muscles supporting the longitudinal arches may run from heel to toes, e.g. flexor digitorum brevis, abductors hallucis and digiti minimi, or be restricted to the forefoot, e.g. lumbricals, interossei, flexor hallucis brevis. Extrinsic muscles lift and support these arches from above, e.g. tibialis anterior and posterior, peroneus longus and brevis.

The *transverse arch* varies in height (as shown) and is sustained by ligaments and muscles, particularly lumbricals, interossei and adductor hallucis.

Flat foot or *pes planus* results from a lowering of the longitudinal arches which may be congenital or the result of muscle weakness and consequent overstretching of supporting ligaments.

Pes cavus is where the medial longitudinal arch in particular is increased in height which again may be congenital or the result of increased muscle tone.

Flattening of the anterior transverse arch may compress plantar digital nerves resulting in pain (*metatarsalgia*).

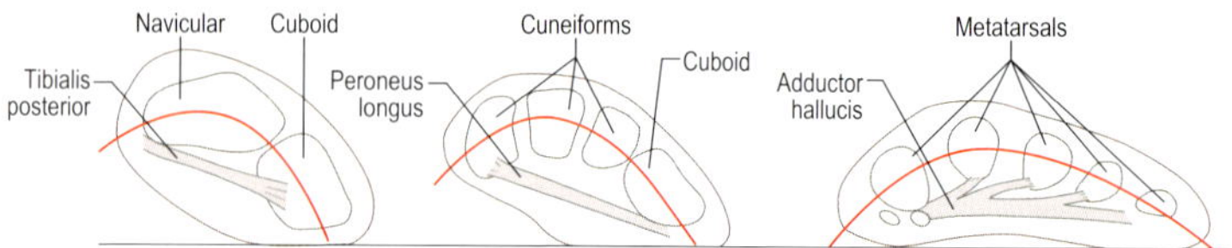

Transverse arches of foot at different levels showing muscles involved in their maintenance

SUMMARY – MUSCLES AND MOVEMENTS

HIP

Extension
Gluteus maximus
All three 'hamstrings'

Abduction
Gluteus maximus
Gluteus medius
Gluteus minimus
Tensor fascia lata

Adduction
Adductors magnus, longus and brevis
Gracilis
Pectineus

Flexion
Psoas major
Iliacus
Pectineus
Rectus femoris
Sartorius

Lateral rotation
Gluteus maximus
Piriformis
Obturators internus and externus
Gemelli
Quadratus femoris

Medial rotation
Gluteus medius and minimus
Tensor fascia lata

KNEE

Flexion
All three 'hamstrings'
Gastrocnemius
Gracilis
Sartorius

Extension
Quadriceps femoris
Tensor fascia lata

Lateral rotation
Biceps femoris

Medial rotation
Semitendinosus
Semimembranosus
Gracilis
Sartorius
Popliteus

ANKLE AND FOOT

Plantarflexion
Gastrocnemius
Soleus
Plantaris
Peroneus longus and brevis
Tibialis posterior
Flexor digitorum longus*
Flexor hallucis longus*

Dorsiflexion
Tibialis anterior
Peroneus tertius
Extensor digitorum longus*
Extensor hallucis longus*

Inversion
Tibialis posterior
Tibialis anterior

Eversion
Peroneus longus, brevis and tertius

TOES

Extension
Extensor hallucis longus (IP, MTP)
Extensor digitorum longus (DIP, PIP, MTP)
Extensor digitorum brevis (DIP, PIP, MTP)
Lumbricals (DIP, PIP)

Flexion
Flexor hallucis longus (IP, MTP)
Flexor hallucis brevis (MTP)
Flexor digitorum longus (DIP, PIP, MTP)
Flexor digitorum brevis (PIP, MTP)
Flexor accessorius (DIP, PIP, MTP)
Flexor digiti minimi brevis (MTP)
Interossei (MTP)
Lumbricals (MTP)

Abduction
Abductor hallucis
Abductor digiti minimi
Dorsal interossei

Adduction
Adductor hallucis
Plantar interossei

*In continued action after moving toes
DIP – distal interphalangeal; IP – interphalangeal; MTP – metatarsophalangeal; PIP – proximal interphalangeal

NERVE SUPPLY

LUMBAR PLEXUS

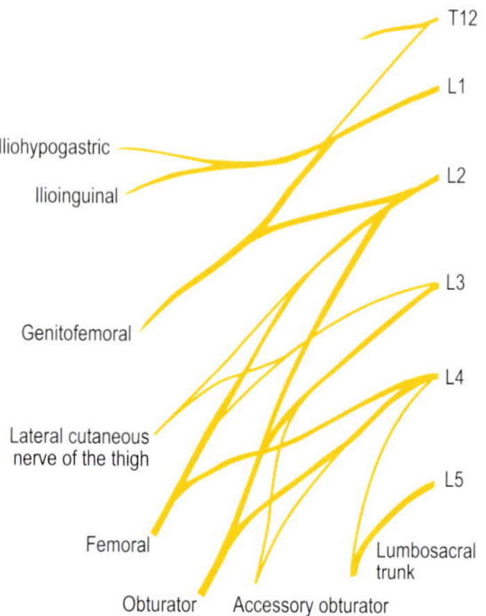

Schematic representation of plexus formation and named nerves

LUMBOSACRAL PLEXUS

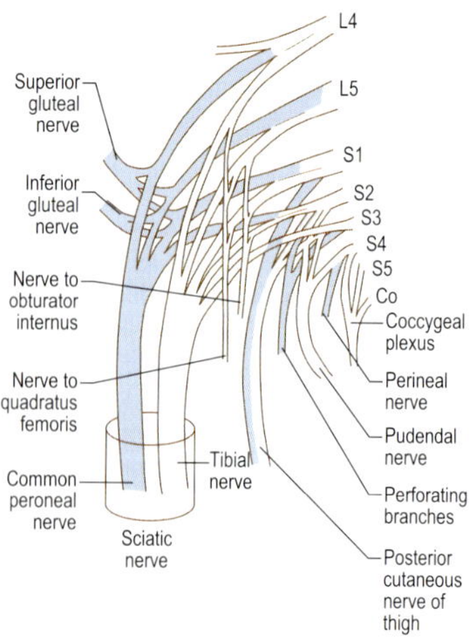

Schematic representation of plexus formation

FEMORAL AND OBTURATOR NERVES

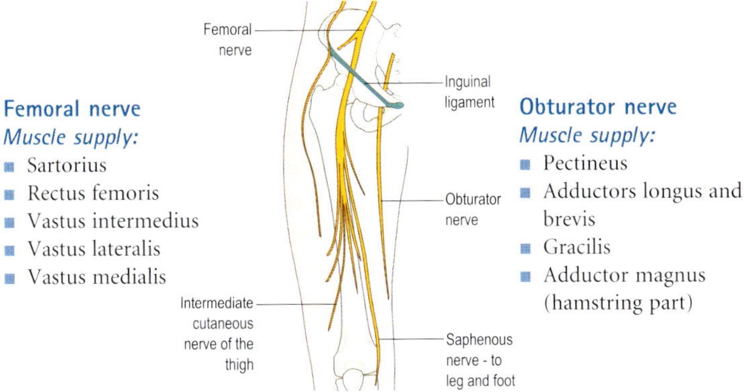

Femoral nerve
Muscle supply:
- Sartorius
- Rectus femoris
- Vastus intermedius
- Vastus lateralis
- Vastus medialis

Obturator nerve
Muscle supply:
- Pectineus
- Adductors longus and brevis
- Gracilis
- Adductor magnus (hamstring part)

Anterior view right thigh

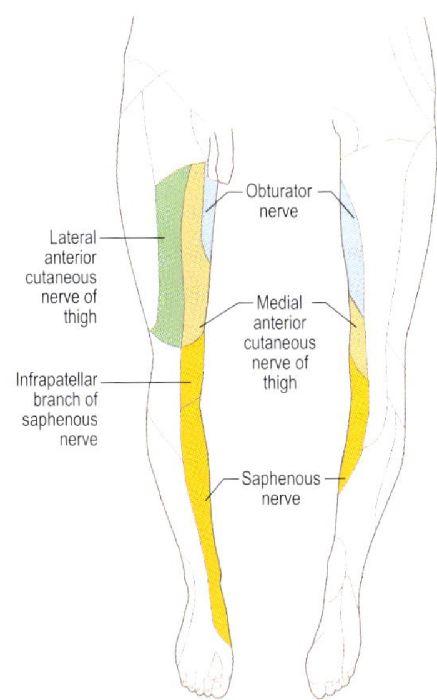

Cutaneous supply: anterior and posterior views of right leg

Obturator nerve
Origin: lumbar plexus, anterior divisions of L2, 3, 4.
Course: formed within substance of psoas major, emerging from its medial border; crosses sacroiliac joint to enter obturator canal below superior pubic ramus.

Femoral nerve
Origin: lumbar plexus, posterior divisions of L2, 3, 4.
Course: formed within substance of psoas major, emerging from its lateral border. Passes into thigh below inguinal ligament entering the femoral triangle where it divides. A major sensory branch, the *saphenous nerve*, is given off just below the inguinal ligament and passes across the thigh to the medial side of knee, continuing along medial side of leg and foot to base of big toe.

Other sensory branches of the femoral nerve are: *lateral anterior* and *medial anterior cutaneous nerves of thigh.*

NERVES OF THE LUMBOSACRAL PLEXUS

The major nerve of the lumbosacral plexus is the sciatic nerve which is the origin of most nerves of the lower limb. Other nerves from this plexus include:

- *superior gluteal nerve* L4, 5, S1 which supplies gluteus medius, minimus and tensor fascia lata
- *inferior gluteal nerve* L5, S1, 2 which supplies gluteus maximus
- *posterior cutaneous nerve of thigh* S1, 2, 3 which is a purely sensory nerve.

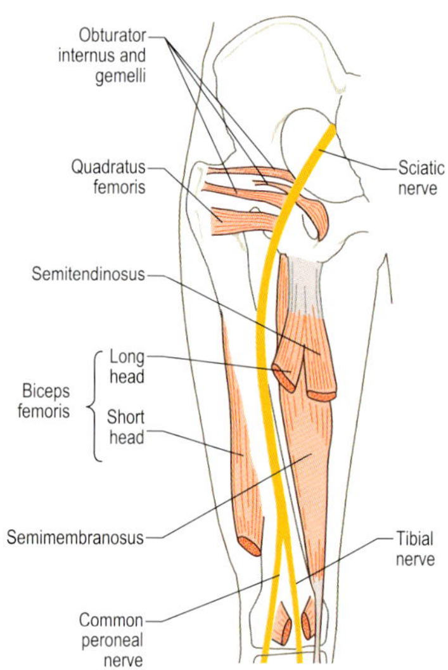

Posterior view left thigh

Sciatic nerve

Origin: ventral rami of L4, 5, S1, 2, 3.
Course: exits pelvis via greater sciatic foramen, below piriformis. Passes down back of thigh deep to biceps femoris, lying on the gemelli, obturator internus, quadratus femoris and adductor magnus. Just below the mid-point of the thigh the sciatic nerve divides into *common peroneal* and *tibial nerves*.
Muscle supply: (via tibial portion of sciatic nerve)

- Semitendinosus
- Semimembranosus
- Biceps femoris (long head)
- Part of adductor magnus

(via common peroneal part of sciatic nerve)

- Biceps femoris (short head)

The sciatic nerve is the largest nerve in the body, measuring up to 1 cm in diameter. Its size is due to the bulk of nerve fibres it contains which will eventually branch into tibial, common peroneal, superficial and deep peroneal, and medial and lateral plantar nerves. Section of the sciatic nerve will therefore result in loss of supply for all these named nerves. Complete section is rare; however, it may be subject to damage at its root origin by pressure or compression at a spinal level. It may also be partially compressed and damaged by a posterior dislocation of the hip.

TIBIAL NERVE, MEDIAL AND LATERAL PLANTAR NERVES

Tibial nerve

Muscle supply

- Gastrocnemius
- Soleus
- Plantaris
- Popliteus
- Tibialis posterior
- Flexor digitorum longus
- Flexor hallucis longus

Lateral plantar nerve

Muscle supply

- Flexors accessorius and digiti minimi brevis
- Abductor digiti minimi
- Adductor hallucis
- Lateral three lumbricals
- Plantar and dorsal interossei

Tibial nerve

Origin: anterior divisions, ventral rami L4, 5, S1–3.

Course: medial terminal branch of sciatic nerve, through popliteal fossa, deep to tendinous arch of soleus. Crosses leg between flexors digitorum and hallucis longus, behind medial malleolus, deep to flexor retinaculum. Divides into *medial* and *lateral plantar nerves* in foot.

Medial plantar nerve

Muscle supply

- Adductor hallucis
- Flexor digitorum brevis
- Flexor hallucis brevis
- 1st (medial) lumbrical

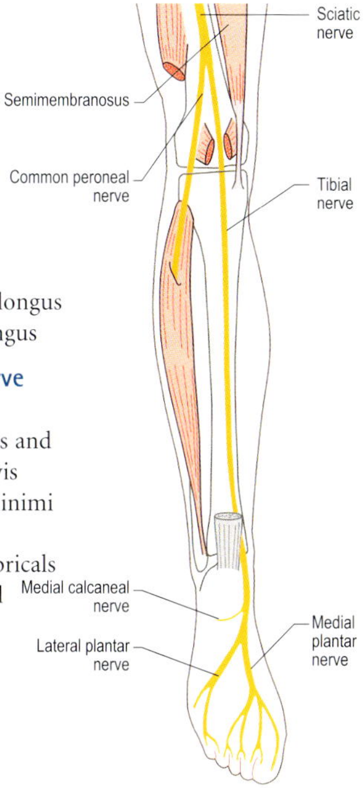

Sciatic nerve
Semimembranosus
Common peroneal nerve
Tibial nerve
Medial calcaneal nerve
Lateral plantar nerve
Medial plantar nerve

Posterior view left leg

Cutaneous supply of tibial nerve

- *Sural nerve* is the major sensory branch of the tibial nerve running from popliteal fossa to base of little toe; it supplies skin posterior and lateral surfaces of leg, lateral border foot and little toe
- *Medial calcaneal nerve* supplies skin over the heel.

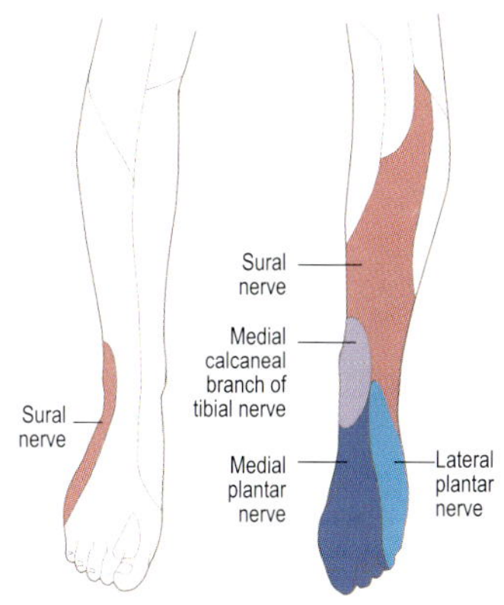

COMMON PERONEAL NERVE

Muscle supply

Superficial peroneal nerve

- Peroneus longus
- Peroneus brevis

Deep peroneal nerve

- Tibialis anterior
- Flexor digitorum longus
- Flexor hallucis longus
- Peroneus tertius
- Extensor digitorum brevis

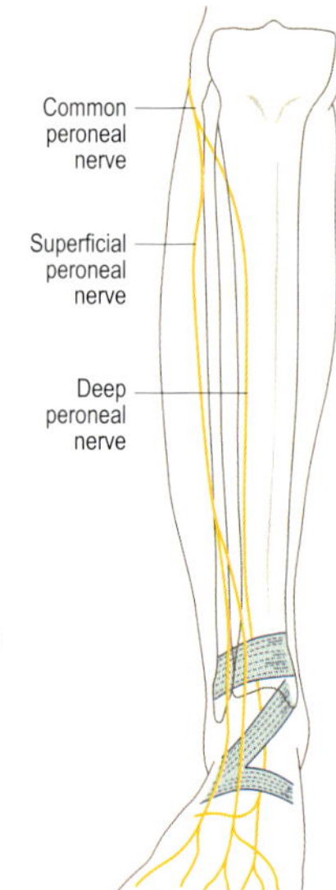

Anterior view right leg

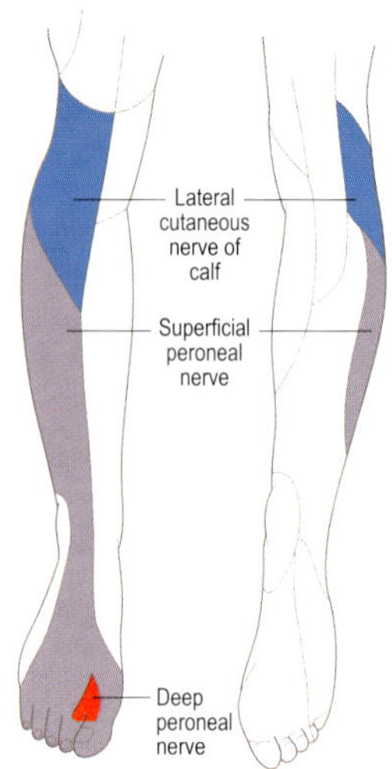

Origin: common peroneal nerve is from posterior divisions L4, 5, S1, 2.
Course: it is the lateral terminal branch of the sciatic nerve crossing the popliteal fossa to pass around the neck of fibula where it divides into *superficial* and *deep peroneal nerves*:

- *superficial peroneal nerve* descends almost vertically along lateral side of leg close to peroneal muscles
- *deep peroneal nerve* passes down in the anterior compartment of leg deep to extensor digitorum longus: it passes below extensor retinacula to enter the dorsum of foot.

Applied anatomy The common peroneal nerve is vulnerable to injury at the neck of fibula where it is commonly crushed by direct trauma (kick, car bumper) or pressure (tight immobilizing cast). This affects both superficial and deep peroneal nerves and the resulting paralysis causes a 'dropped foot', affecting gait. Sensory loss is of less significance.

Cutaneous supply

- *Lateral cutaneous nerve of calf* is the only branch of the common peroneal nerve
- *Superficial peroneal nerve* has an extensive supply over the lateral side of calf and dorsum of foot and toes except the interspace between the hallux and second toe, which is supplied by the *deep peroneal nerve.*

CUTANEOUS NERVE SUPPLY

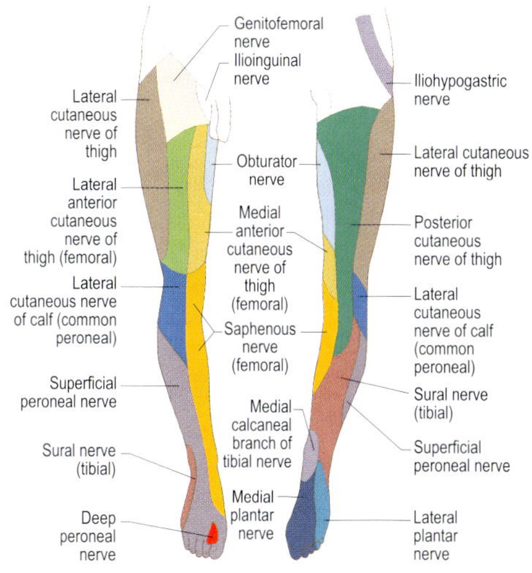

Anterior and posterior views right leg

The *cutaneous distribution* of a named peripheral nerve is the area of skin supplied by this nerve. Severance of the named nerve results in a loss of sensation in the area indicated.

DERMATOMES

A *dermatome* is the area of skin innervated predominantly by one spinal nerve root. This nerve root may have contributed to a number of named peripheral nerves so severance will result in a more variable sensory loss.

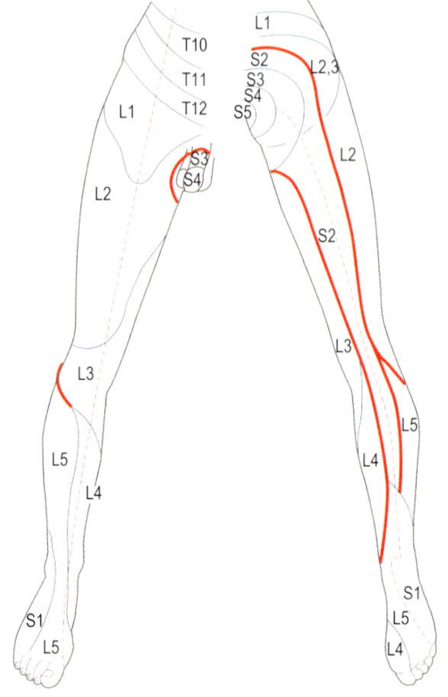

Anterior and posterior views right leg

BLOOD VESSELS

ARTERIES

The main arterial supply to the leg is derived from the *femoral artery* which passes below the inguinal ligament and runs down the thigh to the adductor hiatus where it becomes the *popliteal artery*. The popliteal artery passes through the adductor hiatus and runs vertically through the popliteal fossa, dividing into *anterior* and *posterior tibial arteries* at the lower border of popliteus.

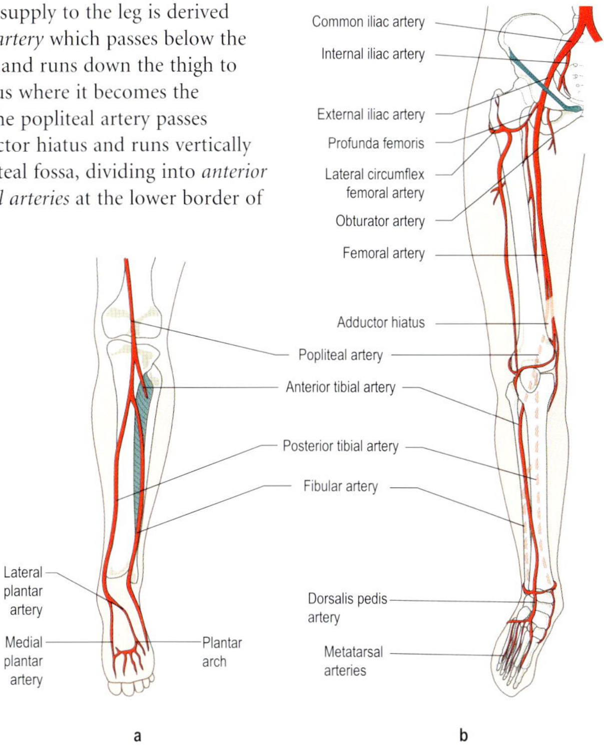

Arteries: (a) posterior and (b) anterior views

The *anterior tibial artery* enters the anterior compartment through the interosseous membrane continuing down its anterior surface, crossing the ankle joint and running between extensors digitorum and hallucis longus (EDL and EHL) tendons, entering the dorsum of foot as the *dorsalis pedis artery*.

The *posterior tibial artery* passes down the back of the leg deep to soleus then runs behind the medial malleolus where it divides into *medial* and *lateral plantar arteries*.

Pulse points The *femoral pulse* can be palpated below the midpoint of the inguinal ligament. The *popliteal pulse* is deep and difficult to find but is in the middle of the popliteal fossa. *Dorsalis pedis pulse* is palpable on the dorsum of the foot between the tendons of EDL and EHL and is the most distal pulse from the heart. The *posterior tibial pulse* is palpated behind the medial malleolus.

VEINS

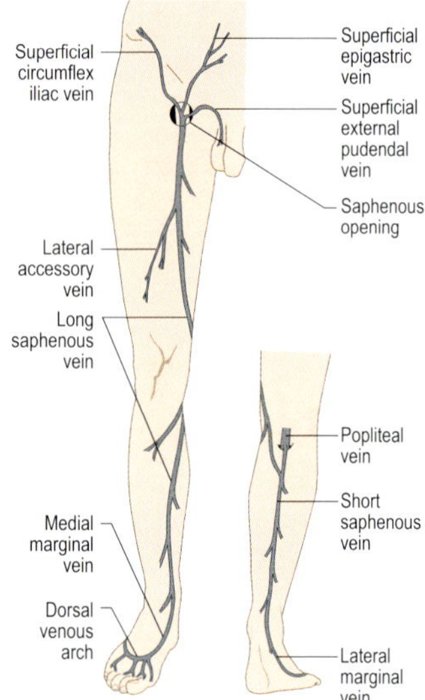

Superficial veins of the lower limb

Veins of the leg are described as *superficial* and *deep* and contain valves only allowing blood flow towards the heart.

The *deep veins* accompany the smaller arteries running in pairs on either side (*venae comitantes*). The popliteal, femoral and profunda femoris veins are single vessels and accompany their named arteries, the femoral becoming the external iliac vein as it passes below the inguinal ligament.

The *superficial veins* are situated in the superficial fascia and have a variable course. The main two are the long (greater) and short (lesser) saphenous veins. The *long saphenous vein* starts as the *medial marginal vein* and passes in front of the medial malleolus, up the calf, behind the medial condyle of the knee, up the thigh, through the saphenous opening to join the *femoral vein*. The *short saphenous vein* starts as the *lateral marginal vein*, passes behind the lateral malleolus, up the posterior aspect of calf and joins the *popliteal vein* behind the knee.

The superficial and deep venous systems are connected by 'perforating veins' and the direction of blood flow allowed by valves is from superficial to deep. If these valves become incompetent flow is reversed, engorging the superficial veins and making them painful and swollen (*varicose veins*). Muscle contraction compresses veins, causing blood flow through the valves towards the heart, an important factor in venous return. The deep veins of the calf are the site for *deep vein thrombosis (DVT)*.

PART 4

The trunk and neck

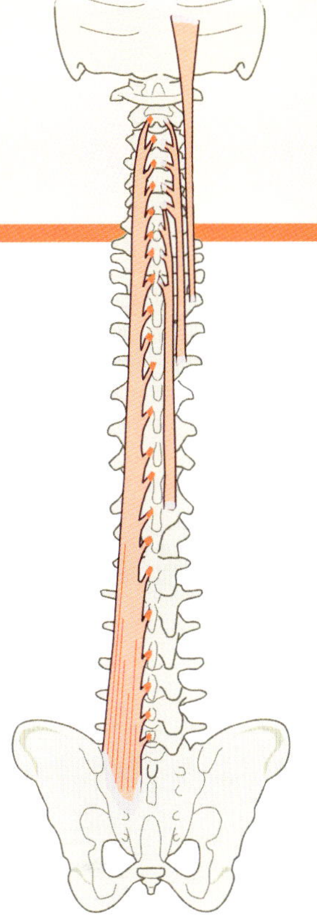

THORAX

INTRODUCTION

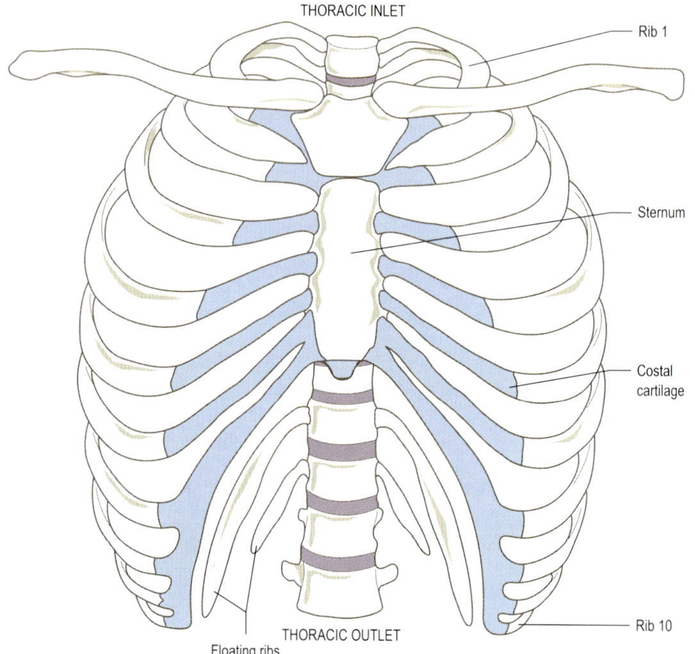

Anterior view thoracic cage

The thoracic cage is a bony and muscular structure which surrounds, protects and supports the heart and lungs amongst other structures. It is narrower towards the neck and wider inferiorly.

The superior *thoracic inlet* is bounded by the body of T1 posteriorly, medial border of the 1st rib and its costal cartilage on both sides, and the superior surface of the manubrium anteriorly. The oesophagus, trachea, vessels and nerves enter or leave the thorax through this inlet.

The inferior *thoracic outlet* is larger and bounded by the body of T12 posteriorly, the 12th and anterior half of the 11th rib on either side, together with the 6th to 10th costal cartilages, and xiphisternal junction anteriorly. The diaphragm covers most of the outlet but has openings for the passage of the aorta (at the level of T12), oesophagus (at the level of T10) and inferior vena cava (at the level of T8). Other smaller structures also pierce the diaphragm to pass between the thorax and abdomen.

RIBS AND STERNUM

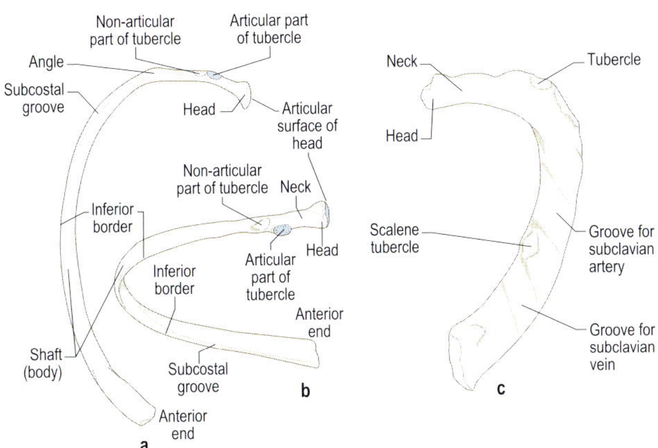

Typical left rib: (a) inferior and (b) posterior views; (c) left 1st rib, superior view

Typical ribs Comprise an enlarged head with two articular facets separated by a ridge, a narrow neck which continues from a tubercle as a flattened curved shaft with rounded upper and sharp lower borders. At the angle the rib turns downwards and inwards. The anterior end is wider, ending as a hollow where it is continuous with its costal cartilage.

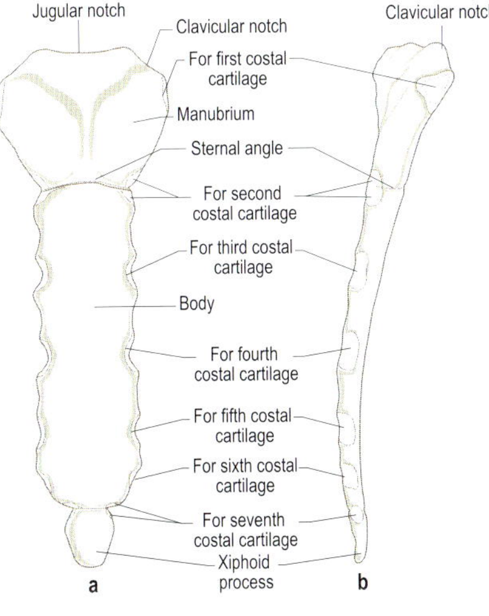

(a) Anterior and (b) lateral views of sternum

First rib Short C-shaped flattened rib with superior and inferior surfaces, and medial and lateral borders. The head has one facet; the neck is relatively long; the large tubercle is on the outer border.

Sternum Elongated flat bone in midline of chest: consists of three parts (superior manubrium, large body, small irregular xiphoid process inferiorly), each with anterior and posterior surfaces and lateral border with facets for costal cartilages. Adjacent parts joined by secondary cartilaginous joints.

Jugular notch is level with T2, sternal angle with T4, xiphisternal junction with T9.

JOINTS

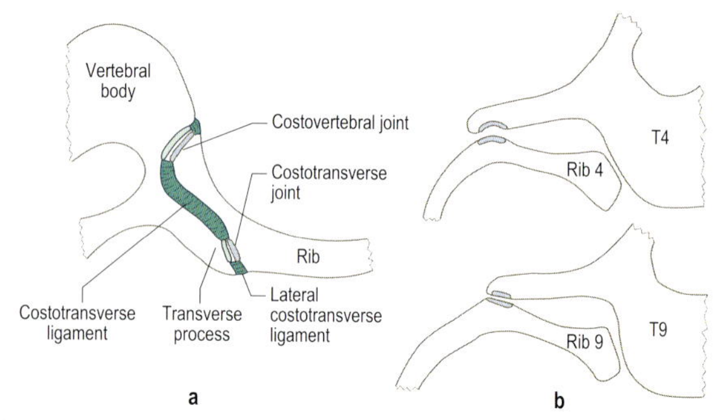

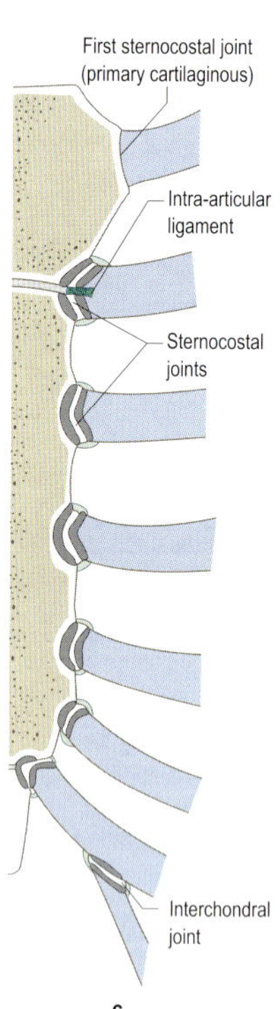

Costovertebral joints Synovial joints between convex articular facets on rib head and concave demifacets on upper border of corresponding vertebra and lower border of vertebra above. The rib head crest articulates with the intervening intervertebral disc. A loose fibrous capsule surrounds the joint thickened anteriorly, forming the *radiate ligament* of the head of rib, blending with the posterior longitudinal ligament. An intra-articular ligament divides the joint space into two.

Costotransverse joints Synovial joints between the tubercles of ribs 1 to 10 and corresponding transverse processes. A thin fibrous capsule surrounds the joint strengthened by, but separated from, the lateral costotransverse ligament posterolaterally.

Sternocostal joints Joints between medial ends of costal cartilage of ribs 1 to 7: the 1st is primary cartilaginous and the remainder synovial. Anterior and posterior radiate ligaments pass either side of the joints.

Interchondral joints Synovial joints between the tips of 8th, 9th and 10th costal cartilages and the cartilage above. Fibrous capsules surround each joint.

(a) Articulation between the rib head and vertebra; (b) shape of the costotransverse joints; (c) sternocostal and intercostal joints

LIGAMENTS

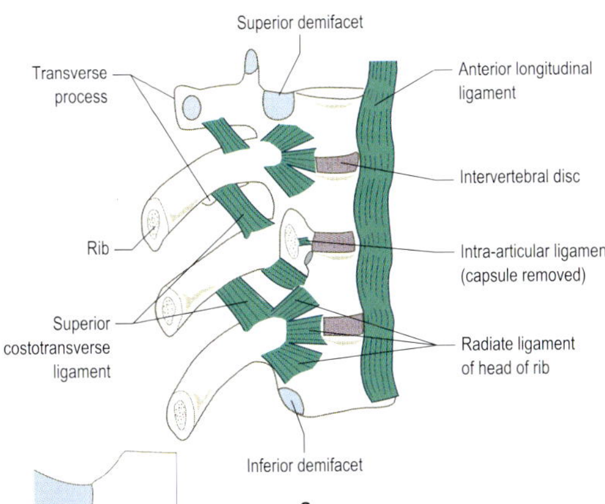

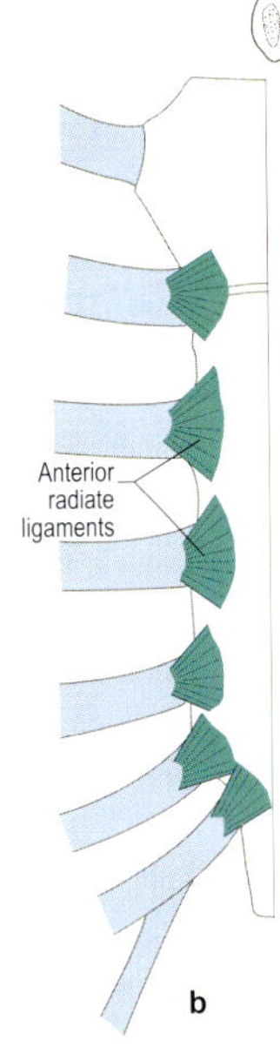

(a) Ligaments associated with the costovertebral, costotransverse and (b) sternocostal joints

Costovertebral joint

Radiate ligament of head of rib – from front of rib head superiorly to vertebral body above, horizontally to front of intervertebral disc and inferiorly to vertebral body below.

Intra-articular ligament – short thick band from crest of rib head to intervertebral disc.

Costotransverse joint

Lateral costotransverse ligament – stout band between tip of transverse process and roughened lateral part of costal tubercle.

Costotransverse ligament – short fibres from back of rib neck to front of transverse process medial to the facet.

Superior costotransverse ligament – two bands from rib to undersurface of transverse process of vertebra above: posterior band passes superomedially, anterior band superolaterally, blending with internal intercostal membrane.

Sternocostal joint

Anterior and posterior radiate ligaments – from medial end of costal cartilage passing superiorly, horizontally and inferiorly to sternum: upper fibres interlacing with those from adjacent and opposite joints.

Intercostal joints

Anterior and posterior oblique ligaments – between adjacent costal cartilages.

MOVEMENTS OF RIBS

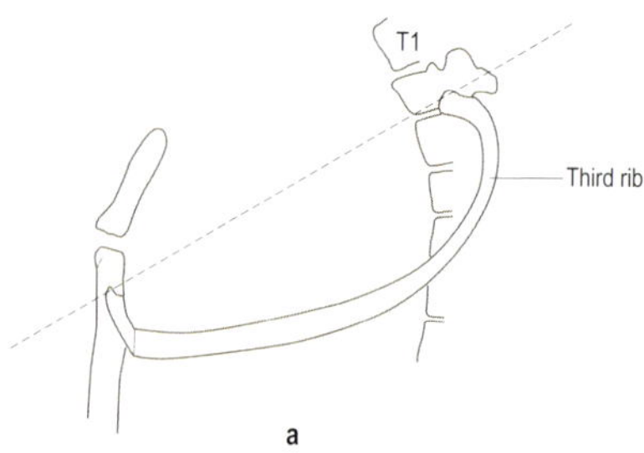

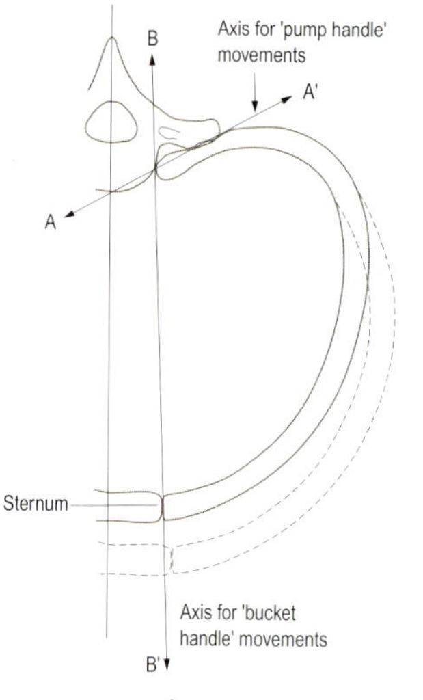

(a) Relationship of anterior and posterior ends of a rib; (b) axes of movement of ribs during respiration

During respiration the anteroposterior, transverse and vertical diameters of the thorax change, resulting in changes in thoracic volume. Vertical diameter changes are due to contraction and relaxation of the diaphragm.

Movement of the 2nd to 5th ribs during inspiration occurs about an axis along the rib neck through the costovertebral and costotransverse joints, causing their anterior ends to be raised ('*pump handle*' movement). Because the 1st rib is firmly attached to the manubrium, movement of its anterior end is slight: the longer 2nd to 5th ribs lift the body of the sternum upwards and forwards, resulting in bending of the manubriosternal joint and an increase in the anteroposterior diameter of the thorax.

In *inspiration* movement of the 8th to 10th ribs occurs about an axis through the costovertebral and sternocostal joints, causing the shaft of the rib to move outwards and upwards ('*bucket handle*' movement), widening the infrasternal angle to increase the transverse diameter of the thorax. Ribs 11 and 12 have little influence on movement but provide firm attachment for the diaphragm.

During *expiration* reverse movements of the ribs and sternum occur, decreasing the anteroposterior and transverse diameters.

RESPIRATORY MUSCLES

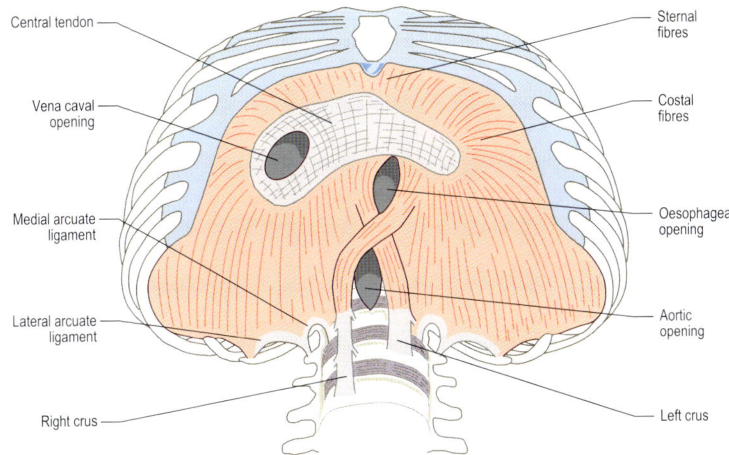

Inferior view of diaphragm

Diaphragm – muscle of inspiration attaching around the thoracic outlet.
Origin: xiphoid process, inner surface lower 6 ribs, lumbar attachment via crura and arcuate ligaments; the right crus is a muscular column arising from bodies L1–3 and the left from L1–2; the medial arcuate ligament runs between body of L2 and transverse process L1, the lateral from transverse process L1 to tip of 12th rib.
Insertion: large, trefoil-shaped central tendon.
Nerve supply: left and right phrenic nerves C3, 4, 5.

The dome of the diaphragm supports two smaller muscular domes (*cupolae*) either side of the central tendon. Inspiration starts with the two cupolae flattening to bring muscle fibres into position to pull down the central tendon (level T8 to T9). The central tendon then becomes fixed and continued muscular contraction lifts the ribs and sternum ('bucket' and 'pump' handle action) to increase thoracic volume, drawing air in.

The aorta (between crura at T12), oesophagus (in right crus at T10) and vena cava (in central tendon T8) pass through the diaphragm.

The high root level of the phrenic nerve means that inspiratory function of the diaphragm can remain unaffected in high spinal cord lesions up to C5 segment.

The diaphragm is 'fixed' during expulsive acts where abdominal muscles contract to raise intra-abdominal pressure whereas abdominal contraction during forced expiration, e.g. coughing, pushes the relaxed diaphragm upwards.

RESPIRATORY MUSCLES

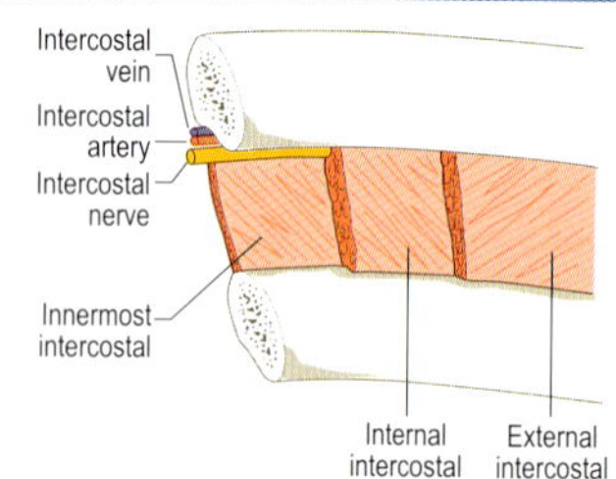

Lateral view intercostal muscles

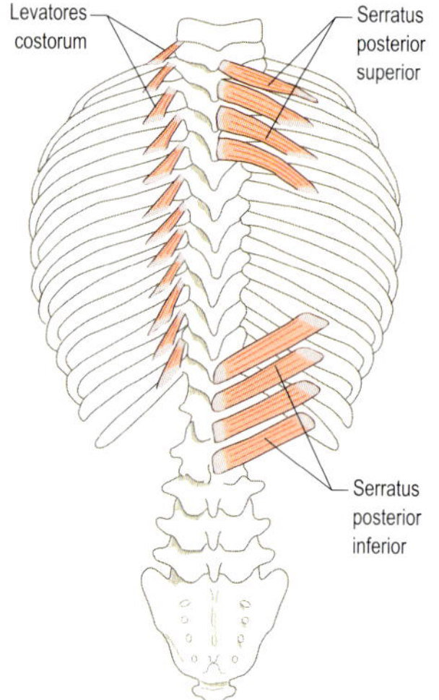

Posterior view thoracic cage

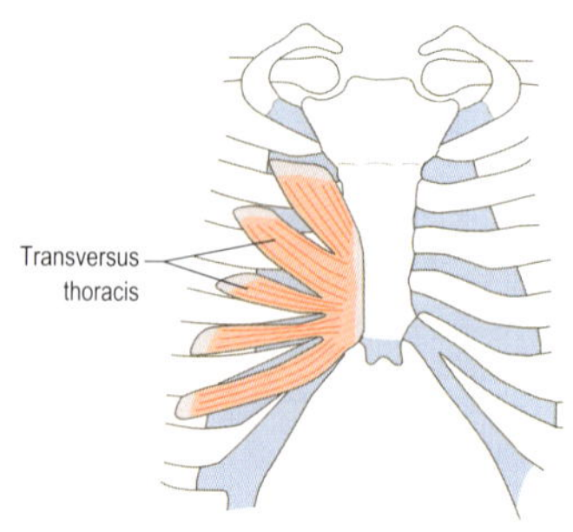

View of inner surface sternum and ribs

Intercostals – external, internal and innermost. External may help lift ribs in inspiration but their main function is to fill the gaps between ribs, creating an airtight cavity. Fibres run obliquely between adjacent ribs, the three layers running at different angles.

Levatores costorum – elevate ribs in inspiration. Twelve small muscles on either side of thorax running between transverse processes and ribs below.

Serratus posterior superior – help inspiration. Lie deep to the rhomboids, arising from the ligamentum nuchae and spinous processes of C7 to T3. Insert into the 2nd to 5th ribs just past their angles.

Serratus posterior inferior – assist expiration. Lie deep to latissimus dorsi arising from spinous processes T11 to L2 and attach to the lower four ribs at their angles.

Transversus thoracis – assist expiration. Lie on the inner surface of the anterior thoracic wall, running from the posterior surface xiphoid process, lower half body of sternum and 4th to 7th costal cartilages. They insert into inner surfaces of 2nd to 6th costal cartilages.

Nerve supply: all of the above muscles are supplied by adjacent spinal nerves.

SURFACE MARKINGS AND RELATIONS OF HEART

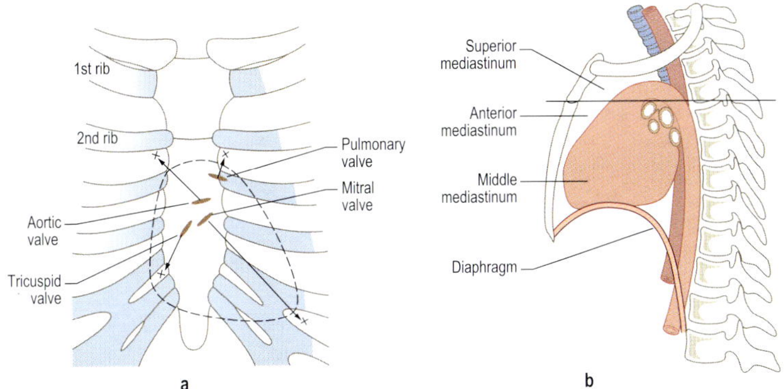

(a) Surface markings of the heart ('x' indicates the site where a stethoscope should be placed to listen to the named heart valve); (b) heart within the middle mediastinum

One-third of the heart lies to the right of midline with the right border extending from the 3rd to 6th costal cartilages, 1 cm from the sternal border. The left border slopes down from the left 2nd intercostal space, 1 cm from the sternal border, to the apex of the heart, whose beat can be felt in the 5th intercostal space, 9 cm from midline. The inferior border lies on the central tendon of the diaphragm: the superior border joins the upper ends of the right and left borders.

The pyramidal-shaped heart has three surfaces: sternocostal (anterior), diaphragmatic (inferior) and base (posterior). The sternocostal surface is formed by the right atrium and ventricle. The diaphragmatic surface is formed by right and left ventricles, the base by the left atrium and small part of the right atrium.

The heart lies in the middle mediastinum surrounded by a double fold of serous membrane (serous pericardium) contained within a dense connective tissue sac (fibrous pericardium), itself attached to the central tendon of the diaphragm. The outer (parietal) layer of serous pericardium lines the fibrous pericardium, while the inner (visceral) layer adheres to the heart wall: the two layers are continuous at the roots of the great vessels forming the closed pericardial space.

The heart wall is mainly cardiac muscle (myocardium), with the visceral serous pericardium and thin subserous layer of connective tissue forming the epicardium, while the chambers are lined by endocardium.

HEART AND ITS BLOOD SUPPLY

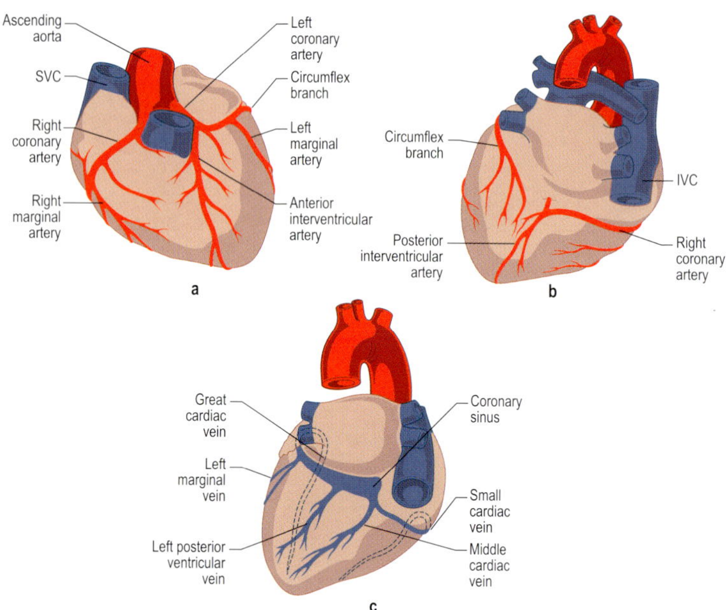

Arterial supply of the heart viewed (a) anteriorly and (b) posteriorly; (c) venous drainage of the heart viewed posteriorly. IVC – inferior vena cava; SVC – superior vena cava

The arterial supply is from branches of the ascending aorta. The *right coronary artery* passes forwards between the pulmonary trunk and right auricle into the atrioventricular groove between right atrium and ventricle, towards the posterior interventricular groove in which it continues as the *posterior interventricular artery*: at the inferior margin of the heart it gives the right marginal branch. The *left coronary artery* passes forwards between the pulmonary trunk and left auricle into the atrioventricular groove between the left atrium and ventricle where it divides into the *circumflex artery*, which continues around the atrioventricular groove, and *anterior interventricular artery*, which runs in the anterior interventricular groove: the circumflex artery gives the *left marginal artery*.

Venous drainage is into the right atrium. The *great cardiac vein* runs in the anterior interventricular groove and on reaching the atrioventricular groove becomes the *coronary sinus*, receiving the *left posterior ventricular* and *left marginal veins*. The *middle cardiac vein* runs in the posterior interventricular groove, while the *small cardiac vein* accompanies the right marginal artery: both drain into the coronary sinus. Anterior cardiac veins drain directly into the right atrium: small veins (venae cordae minimae) drain directly into the atria.

SURFACE MARKINGS OF LUNGS

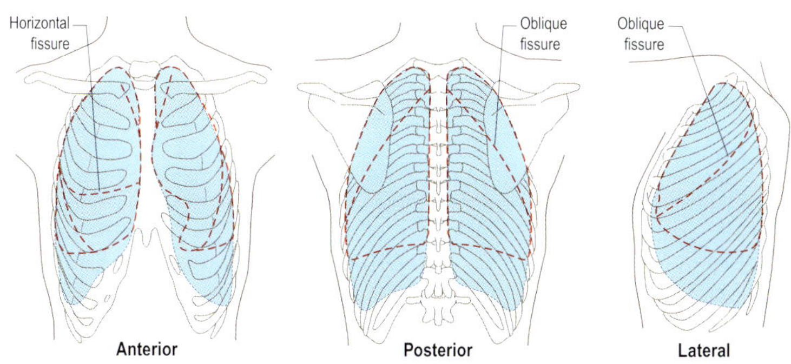

Surface projections of the lungs and visceral pleura (dotted line) and parietal pleura (shaded)

Each lung is surrounded by a pleural sac which has two serous membranous layers (inner visceral, outer parietal) continuous at the lung root. The visceral layer is adherent to lung tissue, extending deep into fissures, while the parietal layer lines the thorax (costal pleura) covering the upper surface of diaphragm (diaphragmatic pleura) and mediastinum (mediastinal pleura).

On both sides, parietal pleura and lung extend 3 cm above the medial third of clavicle, passing behind the sternoclavicular joint to the sternal angle close to the midline. On the right, both pleura run to the 6th costal cartilage where they diverge. On the left, at the 4th costal cartilage, they deviate around the heart before passing to 6th costal cartilage. Passing laterally, the lower border of each lung crosses the midclavicular line at 6th costal cartilage whilst parietal pleura are level with 8th rib. In the midaxillary line the lungs are at 8th rib and parietal pleura at 10th rib, while posteriorly the lungs cross the 10th rib and parietal pleura the 12th rib. Posterior borders of lungs and parietal pleura pass upwards towards apices of the lungs. Posteriorly, the lower borders of the parietal pleura lie below the costal margin. The lung does not extend as far inferiorly as the parietal pleura, creating a potential space, the costodiaphragmatic recess.

The oblique fissure runs from the spinous process of T3 (level with spine of scapula) to the 6th costal cartilage anteriorly: the fissure approximates the medial border of scapula with arm abducted to 90°. The horizontal fissure of the right lung runs laterally along lower border of the 4th rib to meet the oblique fissure in the midaxillary line.

RESPIRATORY TREE

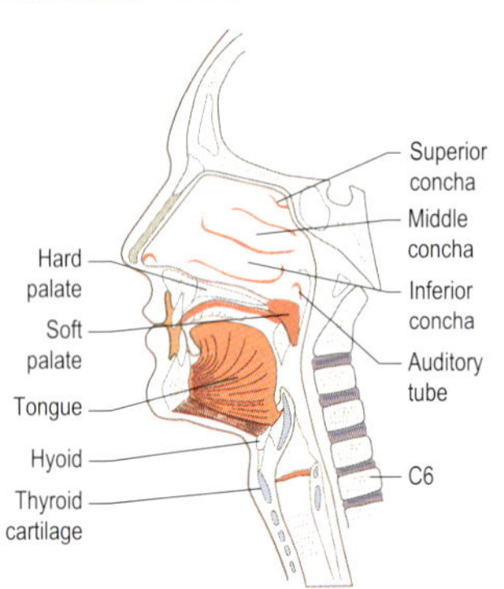

Upper respiratory tract

The respiratory system is concerned with the exchange of gases between air in the lungs and blood in the capillaries.

Upper respiratory tract Rigid or semirigid walls in the nasal cavity, pharynx, larynx, trachea and larger bronchi keep the upper airways open at all times, while the soft flexible lungs respond passively to changes in thoracic diameter. The upper respiratory tract is lined with pseudostratified, ciliated, columnar epithelium (respiratory epithelium) beneath which lies lymphoid tissue, mucus and serous glands and a rich vascular plexus: hence the cleaning, warming and moistening of inspired air.

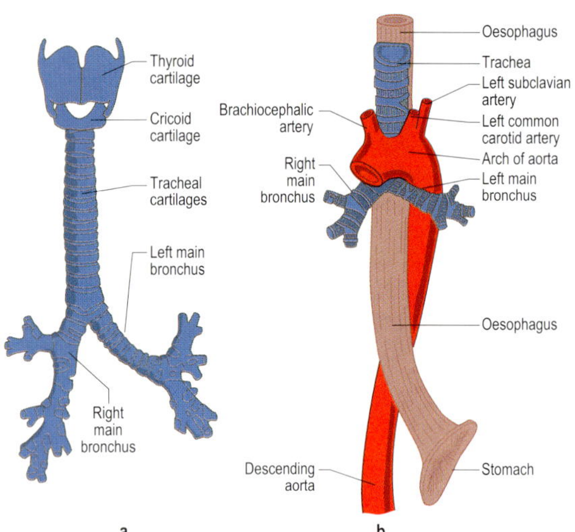

(a) Trachea and principal bronchi; (b) relationship of oesophagus and aorta to trachea

The trachea extends from C6 (cricoid cartilage) to T4/5 where it bifurcates into principal bronchi: the right bronchus is shorter and straighter than the left. The trachea lies anterior to the oesophagus and is crossed by the arch of the aorta above its bifurcation: the descending aorta passes posterior to the left principal bronchus. The principal bronchi divide into secondary bronchi, one for each lobe of the lung, which in turn divide into tertiary bronchi, one for each bronchopulmonary segment.

LUNGS

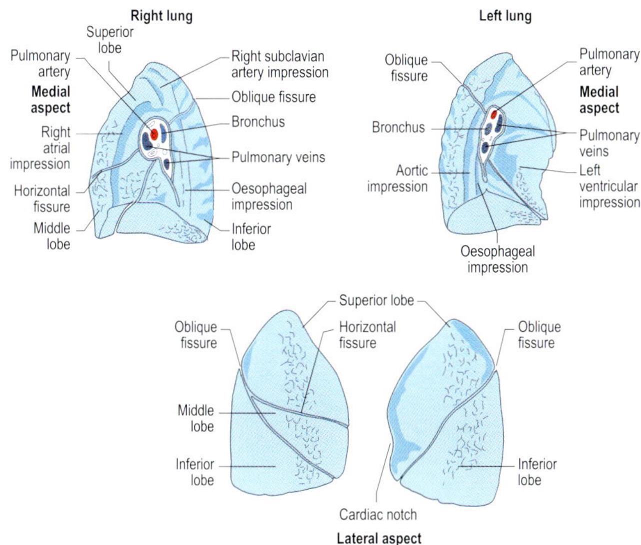

Medial and lateral aspects of right and left lungs

The lungs conform to the shape of the thorax and each has an apex, base, costal and mediastinal surface; an anterior, inferior and posterior border and each attaches to the mediastinum by its root (hilum). The anterior border is sharp, the left having a shallow (cardiac) notch below which is the lingula: the posterior border is rounded and lies in the paravertebral gutter. The shorter, wider and larger right lung is divided into three lobes (superior, middle, inferior) by oblique and horizontal fissures: the left lung has two lobes (superior, inferior) separated by the oblique fissure.

The apex projects into the neck above the junction of the middle and medial third of clavicle: the base is separated from the liver on the right and liver, spleen and stomach on the left by the diaphragm. The medial surface of the right lung is related to the right atrium, oesophagus, trachea and superior vena cava; that of the left to the left ventricle, oesophagus, aortic arch and descending aorta.

The lung root (hilum) contains the bronchus, pulmonary arteries and pulmonary veins: bronchial vessels and lymph nodes are also present. Passing anterior to each hilum is the respective phrenic nerve: posterior is the respective vagus nerve.

VERTEBRAL COLUMN

INTRODUCTION

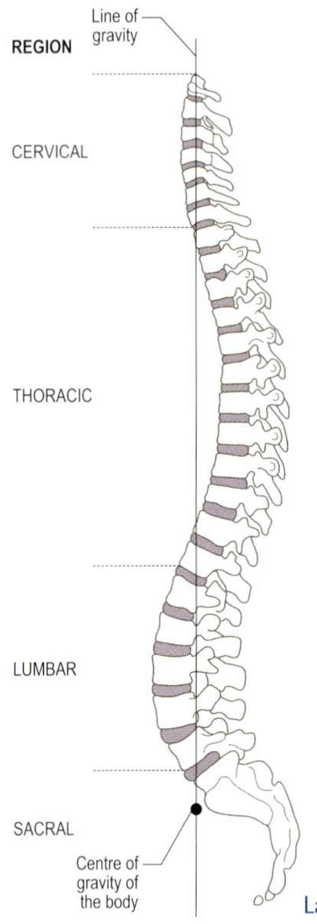

Lateral view vertebral column

The vertebral column (spine) comprises a series of mobile bony segments, each separated by an intervertebral disc and held together by ligaments and muscles. There are 7 cervical, 12 thoracic, 5 lumbar, 5 fused sacral and 4 fused coccygeal vertebrae. Intervertebral discs account for 25% of the length of the vertebral column, which in turn accounts for approximately 40% of an individual's height.

Viewed laterally the adult spine has four curvatures: anterior convexities in the cervical and lumbar regions and anterior concavities in the thoracic and sacral regions. The cervical and lumbar curves are acquired, the lumbar being associated with the development of standing. Viewed anteriorly the spine is straight.

Scoliosis is an abnormal lateral curvature of the spine and compensatory curves develop in the opposite direction to keep the head facing forwards and over the feet. *Kyphosis* is an abnormal anteroposterior curve, usually most marked in the thoracic spine.

Palpation

The most obvious surface markings are the spinous processes:

Lumbar region – L5 can be palpated in a deep hollow just above the sacrum and small gaps between spinous processes allow identification of L4 to T12. Lateral to erector spinae the tips of the transverse processes can be felt on deep palpation.

Thoracic region – spinous processes are easier to identify and posterior aspects of the transverse processes can be felt 2 cm either side of them.

Cervical region – working downwards, C2 is the first to be palpated, with C7 the most prominent. Transverse processes can be palpated lateral to the spinous processes.

CERVICAL VERTEBRAE

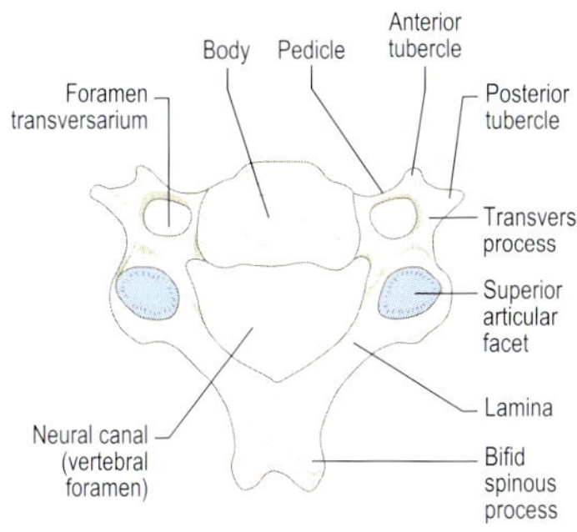

Typical cervical vertebra

Cervical vertebrae have small bodies, with the superior surface projecting upwards at the sides – the inferior surface being correspondingly bevelled. Short pedicles pass posterolaterally and long laminae posteromedially enclosing the triangular vertebral canal. A short bifid spine projects posteriorly from the centre of the vertebral arch. Large articular processes, bearing articular facets, project from the junctions of the pedicles and laminae: superior facets face posterosuperiorly, inferior face anteroinferiorly, both becoming more vertical in the lower cervical region. The transverse process contains the foramen transversarium and ends in anterior and posterior tubercles.

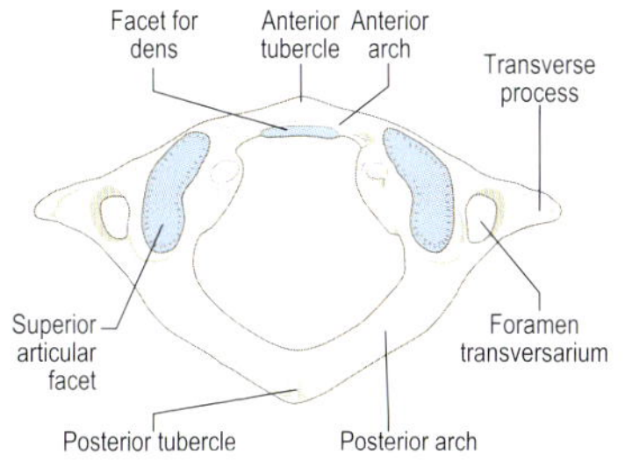

C1 (atlas)

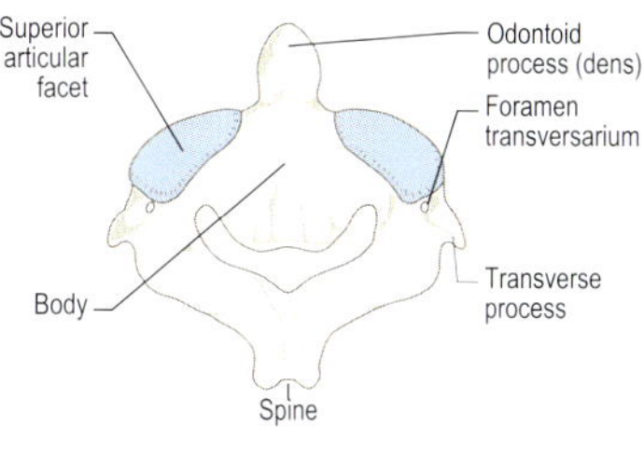

C2 (axis)

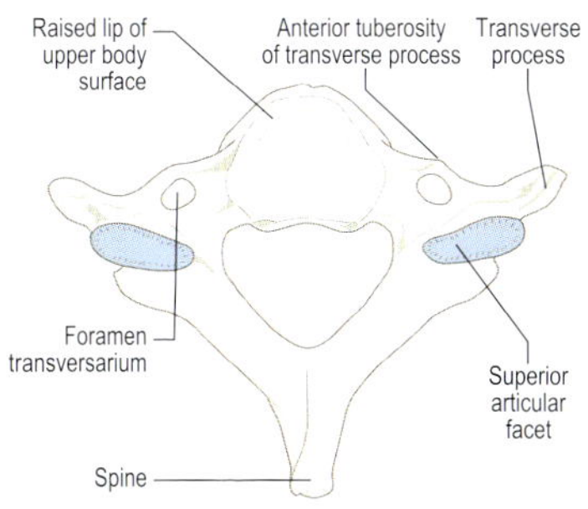

C7

C1 has no body or spine and consists of slender anterior and posterior arches. Concave superior facets articulate with the occipital condyles: the anterior arch articulates with the odontoid process: the inferior articular facets with C2.

C2 has the vertical projecting odontoid process (dens), small round transverse processes and large superior articular facets.

C7 (vertebra prominens) has a long non-bifid spine and shows similarities to thoracic vertebrae.

THORACIC AND LUMBAR VERTEBRAE

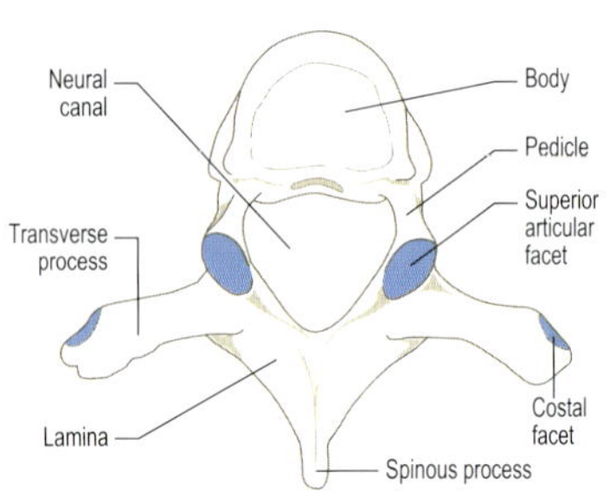

Typical thoracic vertebra

Typical thoracic vertebrae have heart-shaped bodies with articular facets on their sides for articulation with the heads of the ribs. Short pedicles project almost directly posteriorly from the upper part of the body; deep laminae project posteromedially towards the long downward projecting spinous process. Long rounded transverse processes project posterolaterally from the junction of the pedicles and laminae, bearing an articular facet for the tubercle of the rib. Superior and inferior articular processes project almost vertically and have flat articular facets.

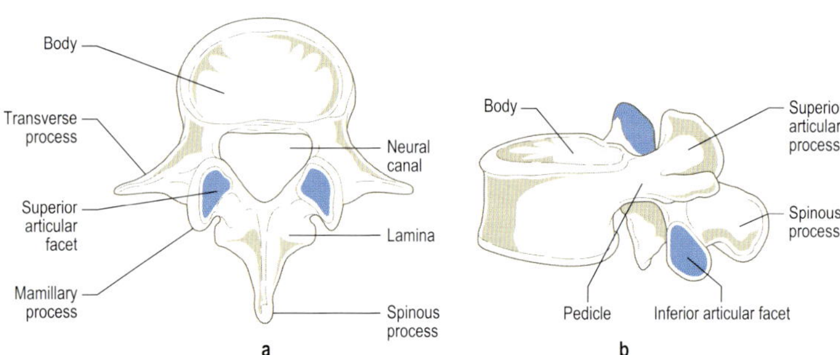

Typical lumbar vertebra: (a) superior and (b) lateral views

Lumbar vertebrae have large bodies with almost parallel superior and inferior surfaces, except for L5 which is deeper anteriorly. Short strong pedicles pass directly posteriorly to join narrow laminae which pass posteromedially towards the horizontal spine. Articular processes project superiorly and inferiorly from the junction of the pedicles and laminae: on the posterior edge of the superior process is the rounded mamillary body. The superior articular facets are concave transversely and flat vertically, the inferior being reciprocally curved. The transverse processes are short and thin.

JOINTS BETWEEN VERTEBRAL BODIES

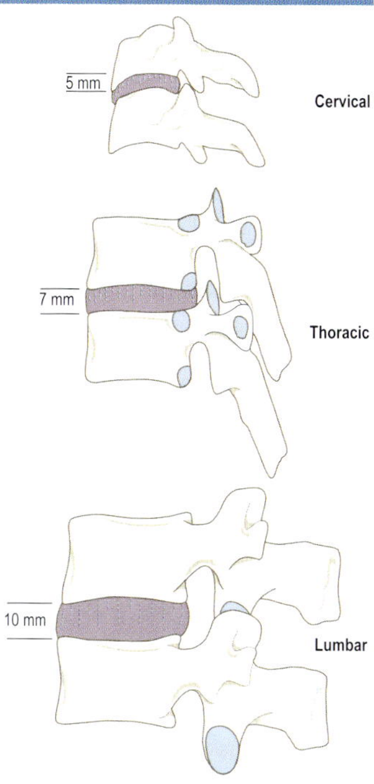

Thickness of the intervertebral disc in different regions of the vertebral column

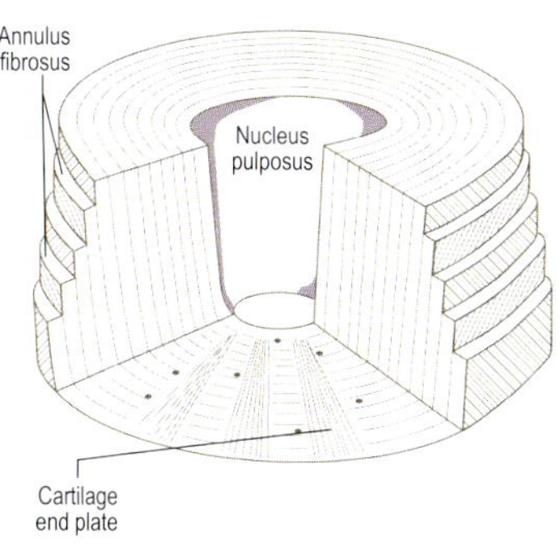

Components of the intervertebral disc

Except for the specialized joints between C1 (atlas) and C2 (axis), joints between bodies of adjacent vertebrae from C2 to S1 are secondary cartilaginous (symphysis). The fibrocartilaginous intervertebral discs are separated from the vertebral body by a thin layer of hyaline cartilage (cartilage end plate).

Intervertebral discs increase in size from above down, averaging 5, 7 and 10 mm in thickness in cervical, thoracic and lumbar regions respectively.

Individual discs are not of uniform thickness, being slightly wedge-shaped to conform to the curvatures of the vertebral column. Each disc consists of three integrated tissues: nucleus pulposus, annulus fibrosus and cartilage end plate.

Nucleus pulposus A three-dimensional lattice of collagen fibres within a hydrophilic proteoglycan gel. Changes in the gel with increasing age lead to dehydration and changes in mechanical behaviour of the disc.

Annulus fibrosus Approximately 20 annular bands surrounding the nucleus, with fibres in each band arranged in alternate directions. This gives great strength and provides restraint to movement.

Cartilage end plate The anatomical limit of the disc to which the annulus attaches. It allows the diffusion of fluid into the disc.

JOINTS BETWEEN VERTEBRAL ARCHES

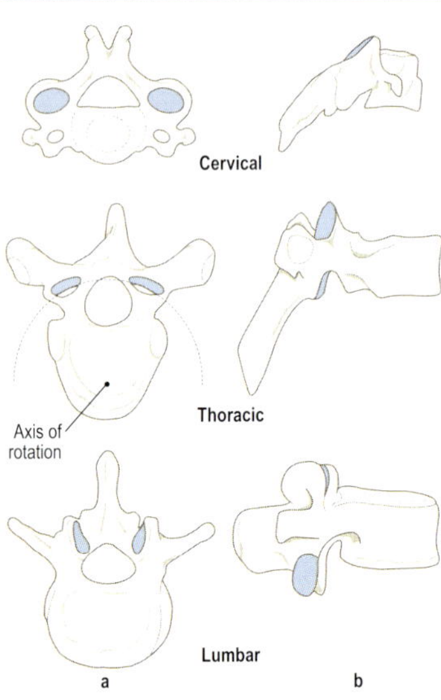

Articular facets of the zygapophyseal joints: (a) superior and (b) lateral views

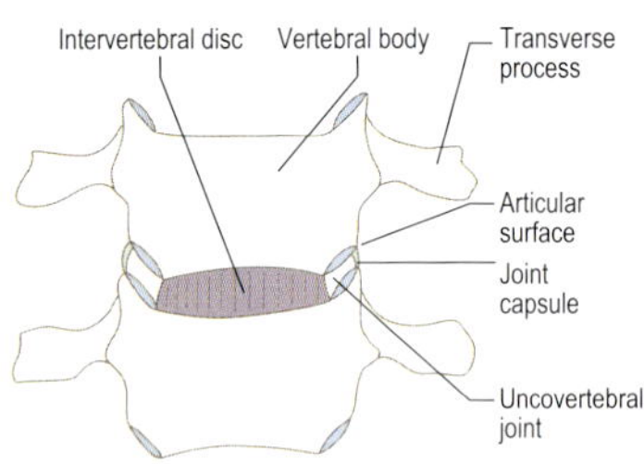

Uncovertebral joints associated with cervical vertebral bodies

Zygapophyseal (facet) joints

Plane synovial joints between facets on the articular processes: their shape and orientation vary in different regions of the spine.

Cervical Joint surfaces are flat oval facets lying in an oblique plane: superior facets face upwards, backwards and medially, inferior facets face downwards, forwards and laterally, the obliquity increasing from above down.

Thoracic Joint surfaces are flat vertically: superior facets face posteriorly with a slight superolateral tilt so that they lie on the circumference of a circle, inferior facets face anteriorly with a slight inferomedial tilt.

Lumbar Facets are reciprocally curved in the horizontal plane and straight in the vertical plane. Superior facets are concave and face posteromedially, interlocking with inferior facets which are convex and face anterolaterally.

Movements: flexion/extension, lateral flexion and rotation, the amount of each depending on spinal level.

Uncovertebral joints Small synovial joints between the lateral parts of adjacent cervical vertebral bodies.

ATLANTO-OCCIPITAL JOINTS

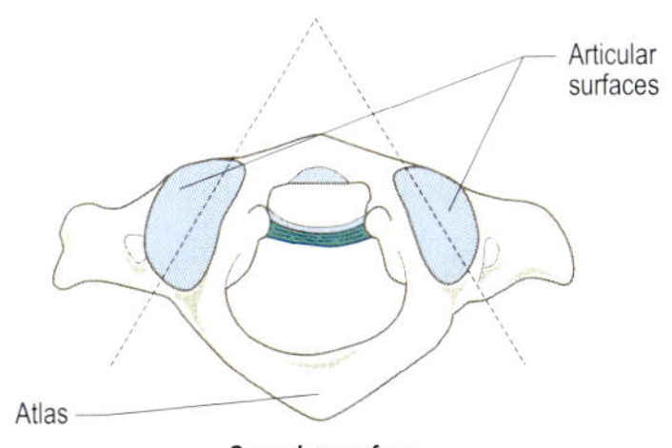

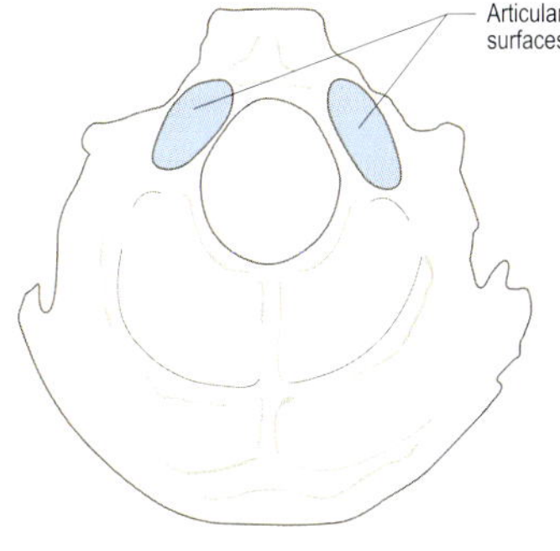

Atlanto-occipital joints Synovial joints between oval concave upper facets of the atlas and the occipital condyles on the base of the skull: the long axes of the joints converge anteriorly.

A thin loose capsule, lined by synovial membrane, attaches to the articular margins of each joint.

Movements: flexion/extension (20°), lateral flexion (8°), limited rotation.

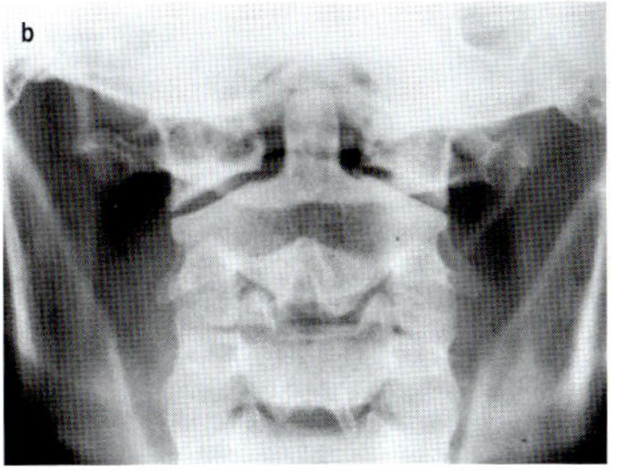

(a) Atlanto-occipital joint surfaces; (b) anteroposterior radiograph of atlanto-occipital and atlantoaxial joints

ATLANTOAXIAL JOINTS

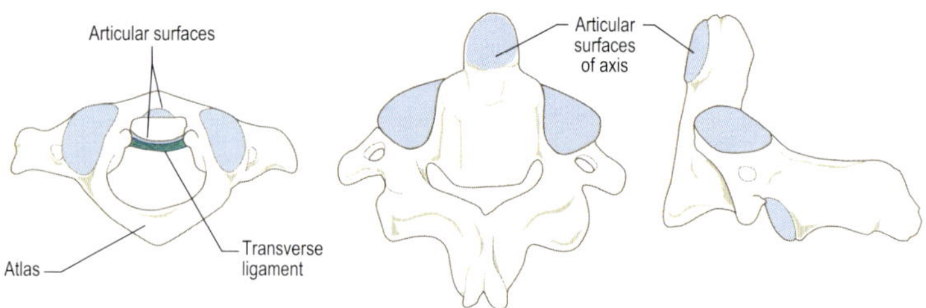

Lateral and medial atlantoaxial joint surfaces

Lateral atlantoaxial joints Synovial ellipsoid joints between flat oval facets on the inferior surface of each lateral mass of the atlas and convex oval facets on the superior surface of the axis. A loose capsule, lined by synovial membrane, attaches to the articular margins of each joint.

Median atlantoaxial joint Synovial pivot joint between the rectangular facet on the front of the dens and an oval facet on the posterior aspect of the anterior arch of the atlas. The posterior surface of the dens articulates with the anterior fibrocartilaginous surface of the transverse ligament of the atlas which completes a fibro-osseous ring. Each joint is enclosed by a thin capsule lined by synovial membrane.

Movements: the main movement at these joints is rotation, with limited flexion/extension.

LIGAMENTS

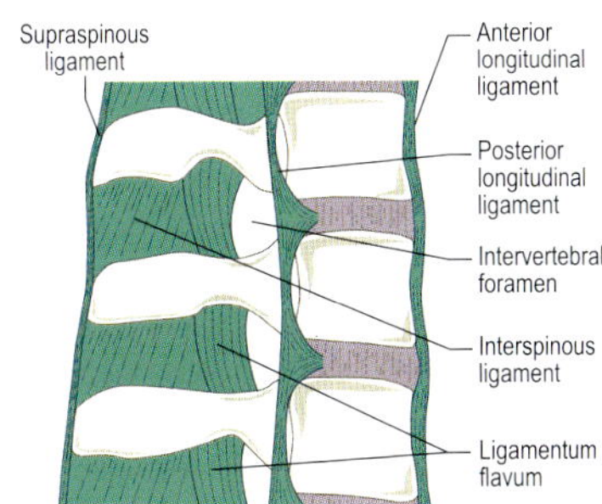

Ligaments of the vertebral column

Anterior longitudinal ligament Attaches to anterior part of bodies and discs from anterior tubercle of atlas to the pelvic surface of sacrum, widening as it descends. It is between 1 and 2 mm thick and consists of three dense layers of collagen fibres.

Posterior longitudinal ligament Attaches to intervertebral discs and adjacent margins of vertebral bodies within the vertebral canal, extending from C2 to sacrum. It is between 1 and 1.4 mm thick and consists of two dense layers of collagen fibres.

Ligamentum flavum Passes between both laminae of adjacent vertebrae from C1 to L5, attaching to front of the lower border of the lamina above and back of the upper border of the lamina below: the medial borders meet at the root of the spine. Contains a large amount of elastic tissue.

Supraspinous ligament Band of longitudinal fibres running over and connecting the tips of spinous processes from C7 to sacrum and is continuous with the posterior edge of the interspinous ligament.

Ligamentum nuchae Triangular midline septum between spine of C7 and external occipital protuberance.

Interspinous ligaments Thin membranous bands between adjacent vertebral spines.

Intertransverse ligaments Pass between adjacent transverse processes: absent in cervical region.

ATLANTO-OCCIPITAL AND ATLANTOAXIAL LIGAMENTS

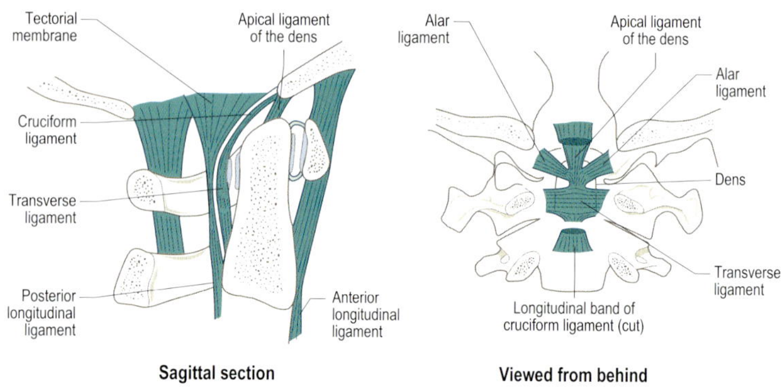

Ligaments associated with atlanto-occipital and atlantoaxial joints (shown in vertical section, with arch removed)

Anterior atlanto-occipital membrane Between anterior arch of atlas and anterior margin of foramen magnum on base of skull, thickened centrally by anterior longitudinal ligament.

Posterior atlanto-occipital membrane Between upper border of posterior arch of atlas and posterior margin of foramen magnum. Its lateral part arches over the vertebral artery and first cervical nerve.

Transverse ligament of atlas Thick strong band attaching to medial aspects of each lateral mass passing behind the dens. Centrally a band of fibres ascends to the anterior margin of the foramen magnum: another band passes downwards to the back of body of the dens. The two longitudinal bands and transverse ligament constitute the *cruciform ligament*.

Tectorial membrane Broad sheet continuous with the posterior longitudinal ligament extending from back of the body of axis to the anterior edge of foramen magnum. Laterally it blends with the accessory atlantoaxial ligament which passes obliquely between the back of atlas and body of axis.

Alar ligaments Short strong round bands passing obliquely anterolaterally from each side of the apex of dens to the medial side of the occipital condyles.

Apical ligament of the dens Slender band passing from the apex of dens to the anterior edge of foramen magnum.

LUMBOSACRAL JOINT AND LIGAMENTS

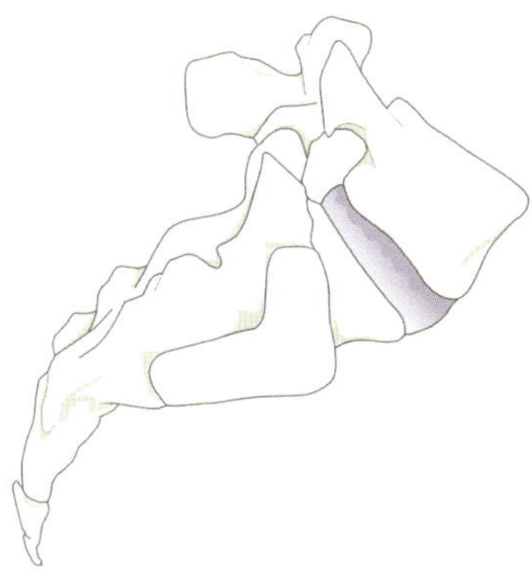

Lumbosacral joint

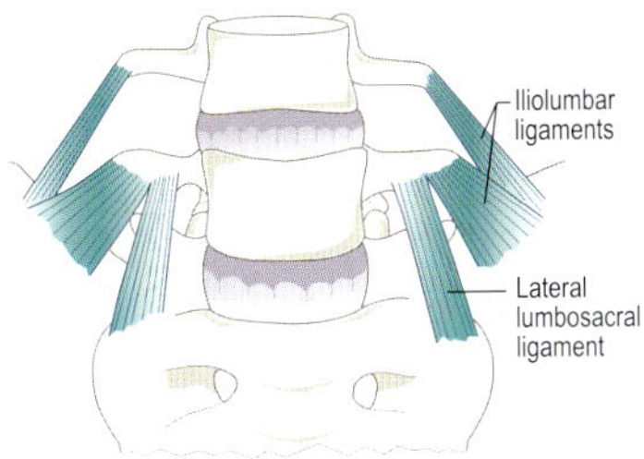

Iliolumbar and lateral lumbosacral ligaments

The lumbosacral joint is between the 5th lumbar and 1st sacral vertebrae. The superior surface of the sacrum is inclined 30° from the horizontal.

The bodies are separated by a wedge-shaped intervertebral disc, while the joints between their articular processes are synovial.

In addition to anterior and posterior longitudinal ligaments, ligamenta flava, interspinous and supraspinous ligaments, two additional ligaments help stabilize the joint:

- *Iliolumbar ligament* passes from the tip of the transverse process of L5 to the posterior part of the inner lip of the iliac crest: a smaller band may pass from the transverse process of L4 to the iliac crest
- *Lateral lumbosacral ligament* passes from lower border of transverse process of L5 to the ala of sacrum.

Movement: flexion, extension and lateral flexion.

> The lumbosacral joint is a weak link in the vertebral column as there is a tendency for L5 to slide forwards because of the inclination of the sacrum. This is resisted by articular processes, disc, spinous ligaments and postvertebral muscles. Fracture of the pars interarticularis allows L5 to slide forward on S1, giving rise to the condition of *spondylolisthesis*.

STABILITY

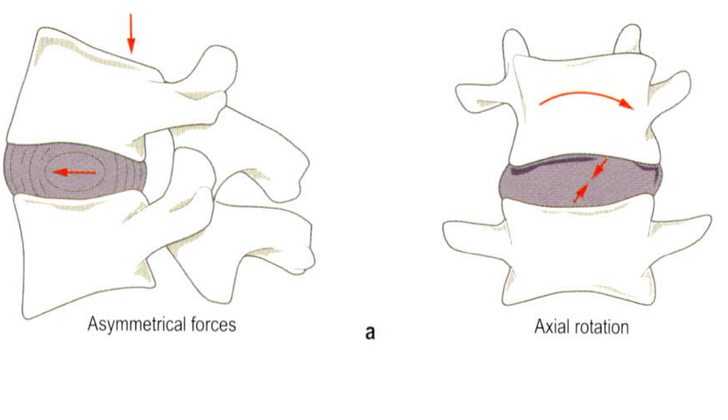

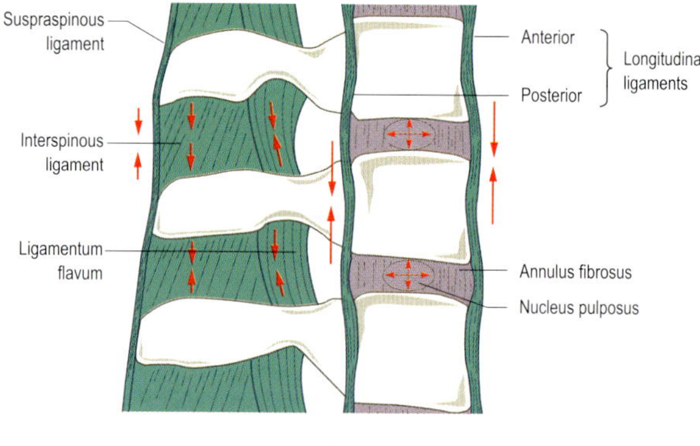

(a) Reaction of the intervertebral disc to asymmetric and axial forces; (b) interaction between disc and ligaments to provide stability

The secondary cartilaginous joints (discs) between vertebral bodies limit the degree of movement between adjacent vertebrae. The annulus fibrosus fibres are constantly under tension due to the preloaded state of the disc, brought about by its water-absorbing capacity.

Asymmetric forces cause the nucleus pulposus to 'move away' from the area of force application, increasing tension in the annulus, tending to restore the upper vertebra to its original position. Axially applied forces stretch annulus fibres counter to the direction of movement compressing the nucleus pulposus, which 'pushes back' against the annulus, limiting movement.

Hydrostatic properties of the disc 'push apart' adjacent vertebral bodies and put all ligaments between vertebrae under tension. This interaction between ligaments and intervertebral discs creates considerable resistance to movement, thereby promoting stability. However, for this arrangement to work efficiently and effectively, the neural arches must be firmly attached to the vertebral bodies.

MOVEMENTS OF THE VERTEBRAL COLUMN

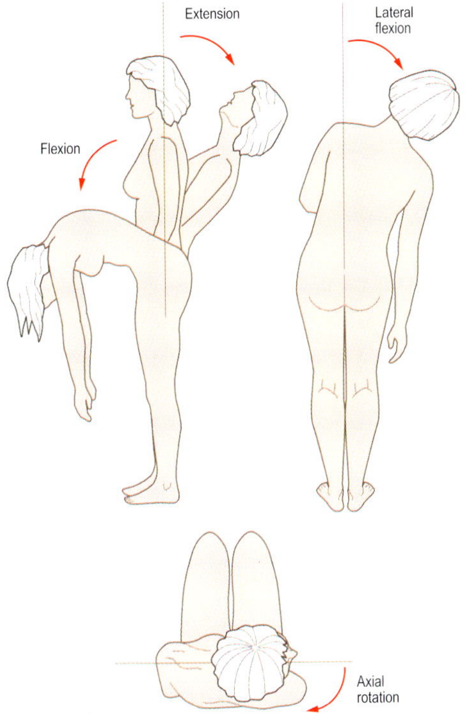

Vertebral column movements

Movement between adjacent vertebrae is due to the slightly flexible intervertebral discs, with the type of movement permitted in each region being largely determined by the shape and orientation of the articular processes. Where intervertebral discs are thick relative to the vertebral body, as in cervical and lumbar regions, movement is increased. Although the degree of movement between successive vertebrae is small, when added together over the whole of the vertebral column the total range becomes considerable.

The basic movements are: flexion (forward bending) and extension (backward bending) about a transverse axis, lateral flexion (bending) about an anteroposterior axis, and rotation to the right or left about a vertical axis. Of these, lateral flexion and rotation are combined movements and neither can occur independently of the other.

MOVEMENTS IN THE SUBOCCIPITAL REGION

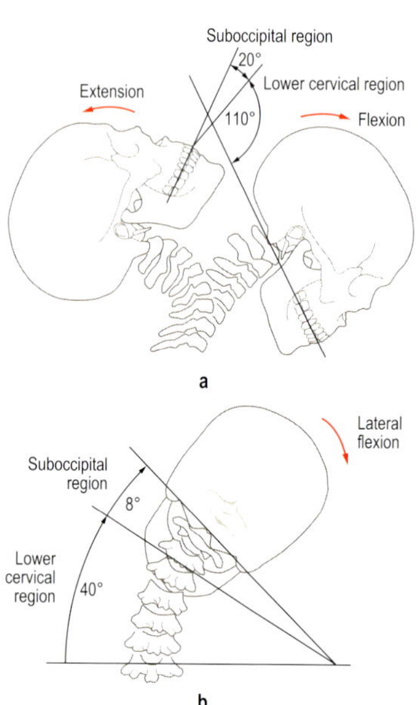

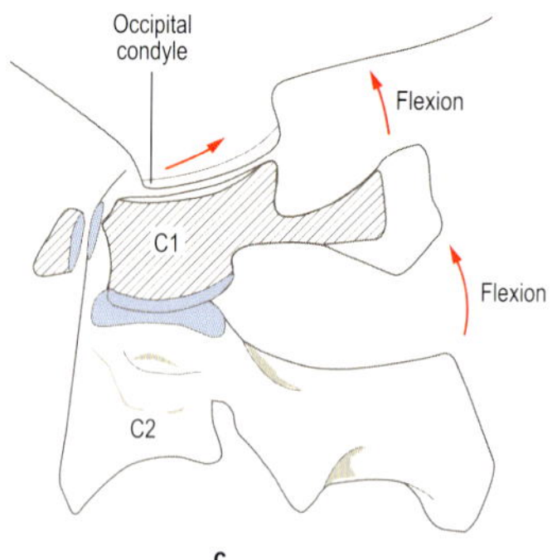

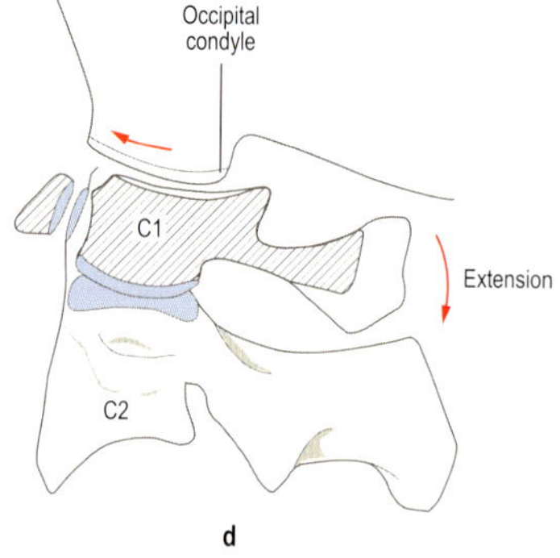

(a) Flexion and extension and (b) lateral flexion of the suboccipital region; (c, d) movement of atlas at atlanto-occipital and atlantoaxial joints during (c) flexion and (d) extension

Flexion and *extension* occur mainly at the atlanto-occipital joints. During flexion and extension the occipital condyles slide posteriorly and anteriorly respectively on the atlas.

Lateral flexion consists of movement of the skull against the atlas (5°) and the axis against the 3rd cervical vertebra (3°).

Axial rotation is rotation of the atlas around the dens: rotation to one side is accompanied by slight lateral flexion to the opposite side.

Accessory movements: The lateral mass of the atlas can be slid forwards against the skull by applying pressure to the posterior arch. The head can also be moved sideways, producing a gliding of one articular surface against the other. Distraction of the joints can be produced by applying traction to the head.

MOVEMENTS OF THE LOWER CERVICAL SPINE

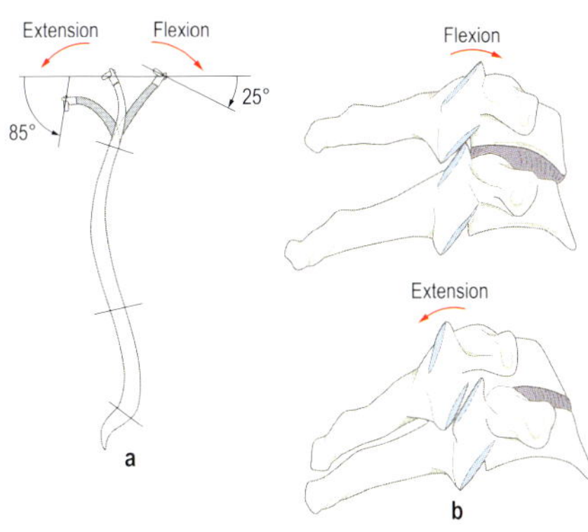

(a) Ranges of flexion/extension; (b) vertebral movements

Flexion and *extension* have a total range of 110°, with extension being greater.

During flexion the upper vertebra tilts and slides anteriorly, its inferior processes moving upwards and forwards compressing the intervertebral space anteriorly and widening it posteriorly. The reverse occurs during extension.

Flexion and extension in the lower cervical region are guided by the uncovertebral joints.

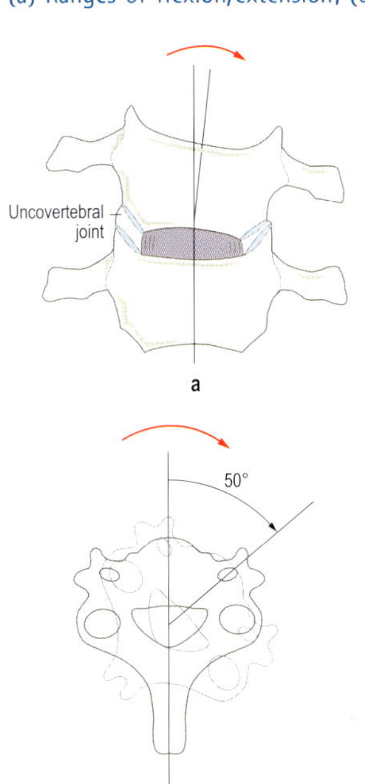

(a) Vertebral movements during lateral flexion; (b) axial rotation

Due to the shape of articular facets lateral flexion and rotation are combined movements.

Lateral flexion has a range of 40° to each side, accompanied by rotation to same side.

Rotation amounts to 50° in each direction. On the side to which rotation is occurring the processes separate, reducing the size of the intervertebral foramen; on the opposite side the inferior process moves against the superior process, increasing the size of the intervertebral foramen.

Accessory movements: local accessory movements can be performed by applying pressure to the spines, transverse processes or articular pillars of the vertebrae. Traction of the neck as a whole draws the nerve roots back through the intervertebral foramina.

MOVEMENTS OF THE THORACIC SPINE

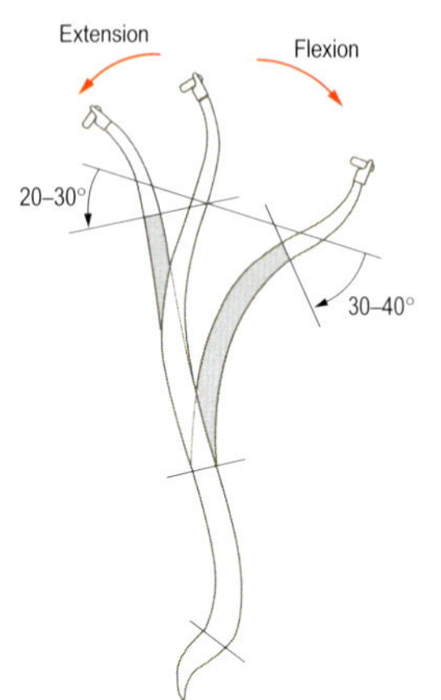

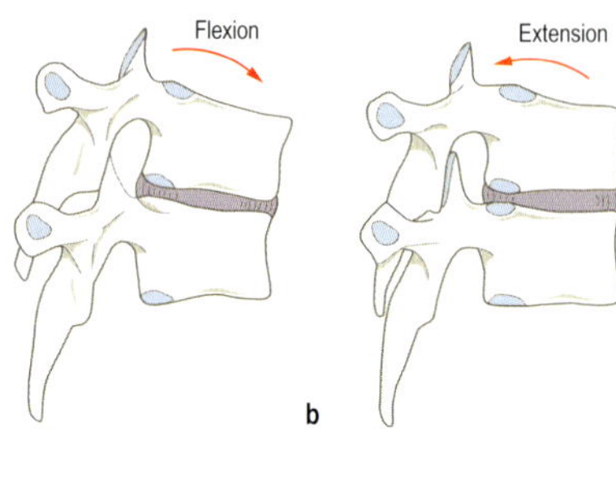

(a) Ranges of flexion/extension; (b) vertebral movements

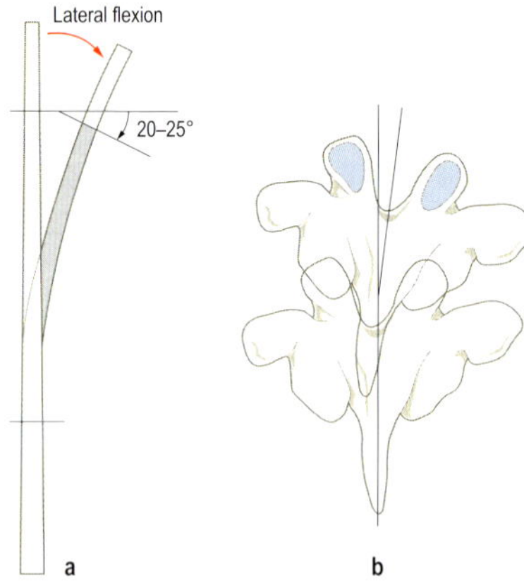

(a) Range of lateral flexion; (b) vertebral movements

The combined range of *flexion and extension* is 50–70°; both movements are limited by the thoracic cage.

In *flexion* the inferior articular process of the vertebra above slides upwards over the superior process of the vertebra below, opening out the interspace posteriorly and compressing the intervertebral disc anteriorly.

In *extension* the vertebrae are approximated posteriorly, being limited by impact of articular and spinous processes.

Lateral flexion is associated with rotation, such that in full lateral flexion (20–25° in each direction) there is 20° of rotation.

In lateral flexion the articular processes of two adjacent vertebrae slide relative to one another, being freer in the lower half of the thoracic spine.

Lateral flexion modifies the shape of the thorax: intercostal spaces widen on the contralateral side and narrow on the ipsilateral side.

Rotation is 35° in each direction, limited by the ribs and facilitated by the direction of the vertebral facets. During walking, rotation of the thoracolumbar region is important.

MOVEMENTS OF THE LUMBAR SPINE

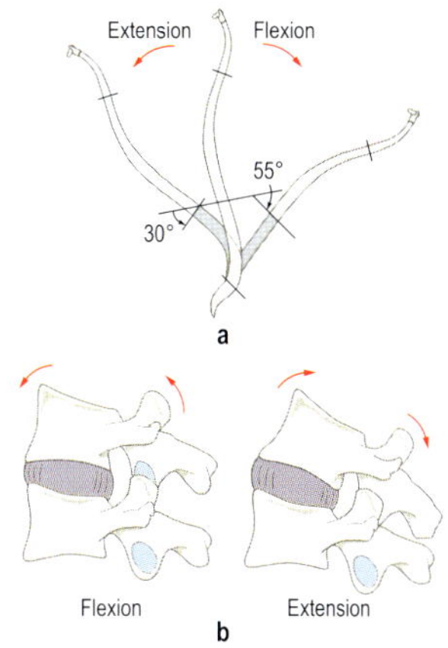

(a) Ranges of flexion/extension; (b) vertebral movements

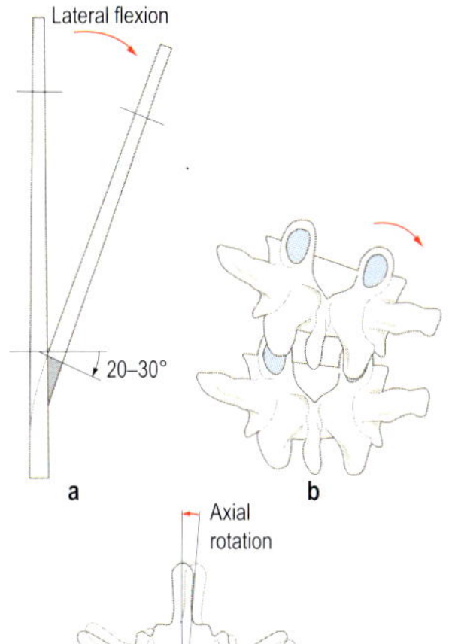

(a) Range of lateral flexion; (b) vertebral movements; (c) axial rotation

Flexion and extension are relatively free, having a total range of 85°. There is less movement at the thoracolumbar junction than between adjacent vertebrae, with most movement occurring at the lumbosacral junction.

During flexion the contact area between articular facets decreases, whereas in extension it increases. There are also changes in the size and shape of the intervertebral foramen, which elongates and narrows during flexion.

Lateral flexion may be as great as 60° to either side in childhood, decreasing to 20–30° in adults. Lateral flexion at the lumbosacral junction is minimal.

Axial rotation is only a few degrees, being limited by the shape and orientation of articular processes.

Accessory movements: application of pressure to one side of a lumbar spinous process elicits a small degree of rotation. Pressure on the mamillary or spinous processes causes anterior gliding of the same vertebra.

RELATIONS AT C6

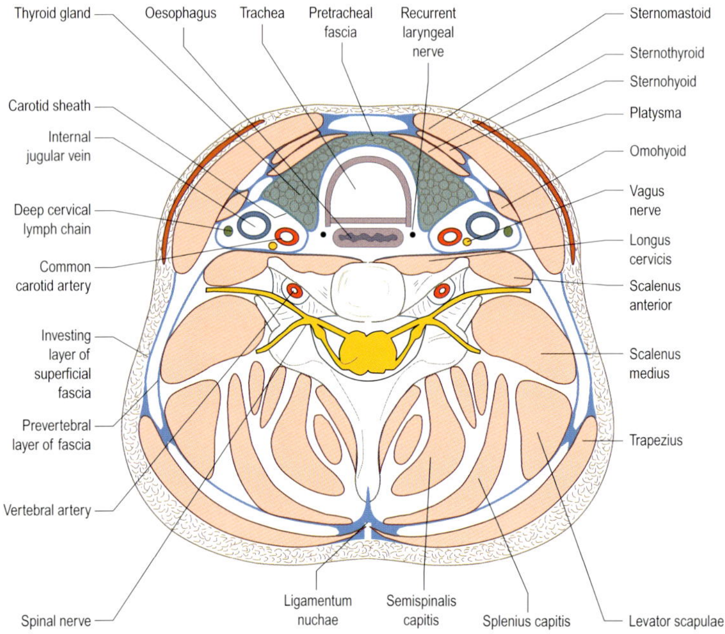

Schematic representation of relations of the vertebral column at the level of C6

In the neck well-defined fascial sheets and membranes enclose the main muscle masses, viscera, main nerves and blood vessels. These fasciae have continuities and bony attachments which form fascial planes and compartments in which deeper structures of the neck are confined.

The postvertebral muscles, enclosed within the superficial layer of cervical fascia, lie posterior to the cervical vertebrae. Anterolaterally is the carotid sheath, enclosing the common carotid artery/internal carotid artery, internal jugular vein, vagus (10th cranial nerve) and the cervical sympathetic trunk. The anterior part of the neck comprises two compartments: a superficial muscular compartment and a deeper visceral compartment enclosing the pharynx, oesophagus, larynx, trachea, thyroid and parathyroid glands.

NECK MUSCLES

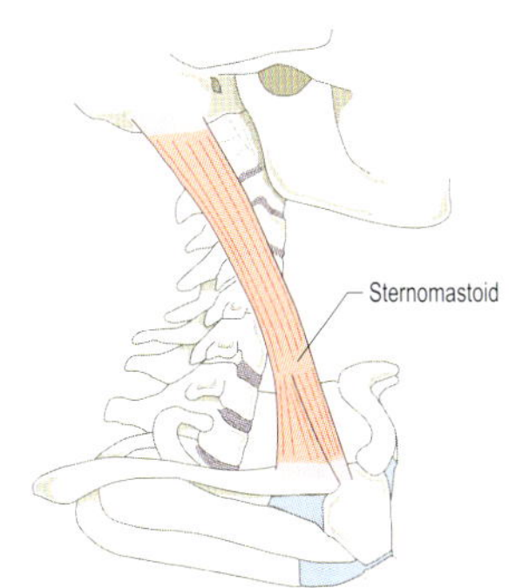

Lateral view right side

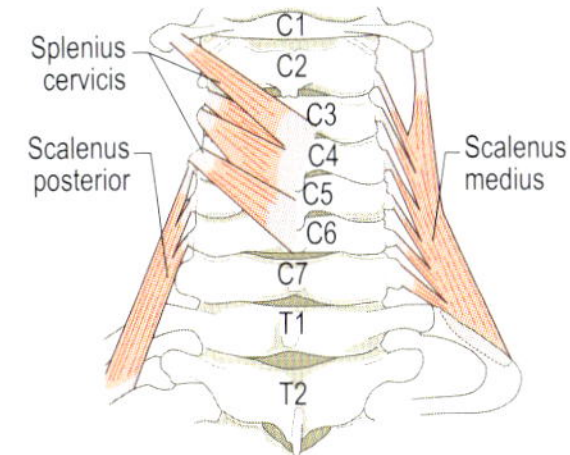

Posterior view

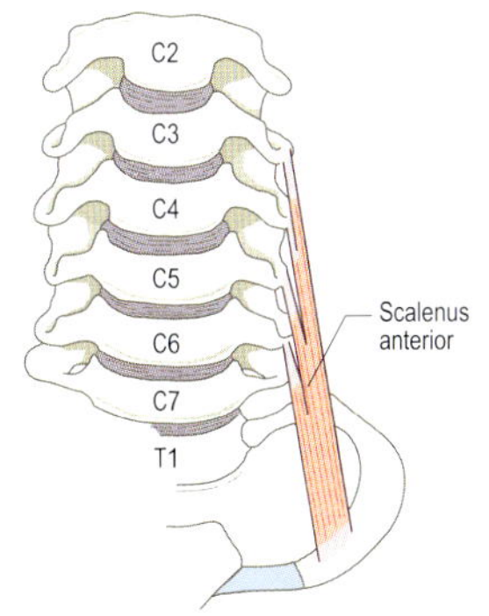

Anterior view

Sternomastoid – both muscles flex neck: one produces lateral flexion to same side and rotation to opposite side.
Origin: sternal head – anterior surface of manubrium; clavicular head – upper surface medial third clavicle.
Insertion: mastoid process and lateral third superior nuchal line.
Nerve supply: cranial accessory nerve (XI); anterior primary rami C2, 3.

Sternomastoid is easily seen and palpated as it passes diagonally across the neck. Spasm of muscle causes *torticollis* where the neck is rotated to opposite side and laterally flexed to same side. Working in reversed origin/insertion it can lift the clavicle and sternum to assist forced inspiration.

Splenius cervicis – both extend neck whilst one laterally flexes and rotates to same side: runs from spinous processes C3–6 to transverse processes C1–4.
Nerve supply: posterior primary rami C5, 6, 7.

Scaleni muscles – all three scaleni laterally flex neck to same side and, working in reversed origin/insertion, lift ribs to assist forced inspiration. The two anterior muscles also flex the neck.

Scalenus anterior
Origin: transverse processes C3–6.
Insertion: scalene tubercle inner border 1st rib.
Nerve supply: anterior primary rami C4, 5, 6.

Scalenus medius
Origin: transverse processes C1–7.
Insertion: upper surface 1st rib.
Nerve supply: anterior primary rami C3–8.

Scalenus posterior
Origin: transverse processes C4–6.
Insertion: outer surface 2nd rib.
Nerve supply: anterior primary rami C6, 7, 8.

NECK MUSCLES

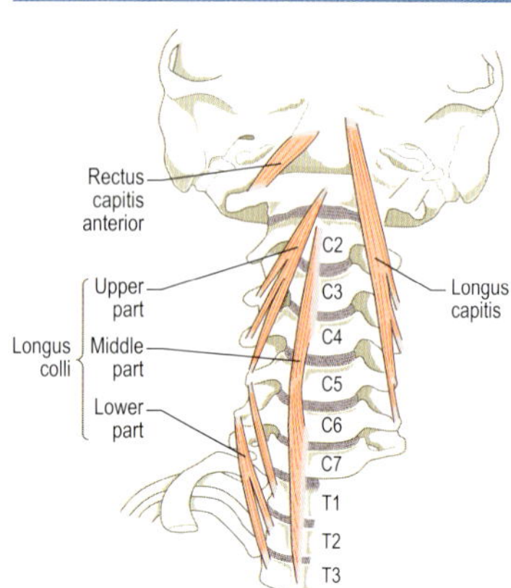

Anterior view cervical spine

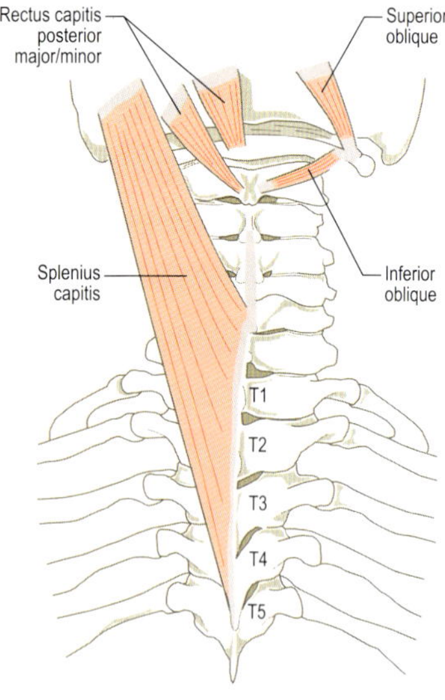

Posterior view cervical spine

Longus colli – flexes neck. In three parts: lower part – bodies T1–3 to transverse processes C5, 6; middle part – bodies C5–T3 to bodies C2–4; upper part – transverse processes C3–5 to anterior tubercle of atlas.
Nerve supply: anterior rami C3, 4, 5, 6.

Longus capitis – flexes neck and head on neck: from transverse processes C3–6 to basilar part occipital bone.
Nerve supply: anterior rami C1, 2, 3 (4).

Rectus capitis anterior – flexes and stabilizes head on neck: from anterior surface lateral mass atlas to basilar part occipital bone.
Nerve supply: anterior rami C1, 2.

Splenius capitis – both sides extend head and neck, one side laterally flexes and rotates face to same side: from ligamentum nuchae and spinous processes C7–T4 to posterior aspect mastoid process and adjacent superior nuchal line.
Nerve supply: posterior rami C3, 4, 5.

Rectus capitis posterior major and minor – both extend head on neck and stabilize atlanto-occipital joint: major – from spinous process of axis to inferior nuchal line; minor – from posterior tubercle atlas to inferior nuchal line.
Nerve supply: posterior ramus C1.

Superior and inferior obliques – superior – from transverse process atlas to inferior nuchal line, extends head on neck; inferior – from spinous process of axis to transverse process of atlas, turns face to same side: both stabilize atlantoaxial joint.
Nerve supply: posterior ramus C1.

BACK MUSCLES

Erector spinae is a large complex muscle consisting of several parts running the length of the vertebral column. It has a broad lumbosacral origin which divides into three columns as it extends upwards, each column further dividing into three parts.

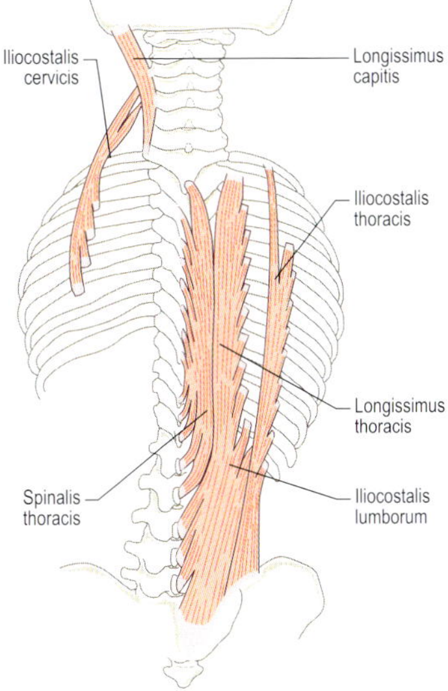

Posterior view erector spinae

Origin: spinous processes and supraspinous ligaments T11–L5, posterior sacral crests, sacrotuberous ligament and posterior part iliac crest.

Iliocostalis – lateral part divided into: *lumborum* – attaches to lower 6 ribs; *thoracis* – lower 6 to upper 6 ribs; *cervicis* – upper 6 ribs to transverse processes C4–7.

Longissimus – intermediate part divided into: *thoracis* – transverse processes of lumbar vertebrae to transverse processes all thoracic vertebrae and adjacent part lower 10 ribs; *cervicis* – transverse processes T1–6 to those of C2–6; *capitis* – transverse processes T1–5 and articular processes C4–7 to mastoid process.

Spinalis – relatively insignificant medial part divided into thoracis, cervicis and capitis, the latter two frequently blending with adjacent muscles: *thoracis* – spinous processes T11–L2 to those of T1–6.

Nerve supply: adjacent posterior rami.

All three columns of erector spinae working together on both sides extend lumbar, thoracic, cervical spines and head on neck; working on one side only, they produce lateral flexion and rotation to same side.

Functionally, erector spinae works to maintain upright human posture. Standing on one leg erector spinae on the non-weight-bearing side helps prevent the pelvis dropping.

Erector spinae can be easily palpated on either side of the lumbar spine.

BACK MUSCLES

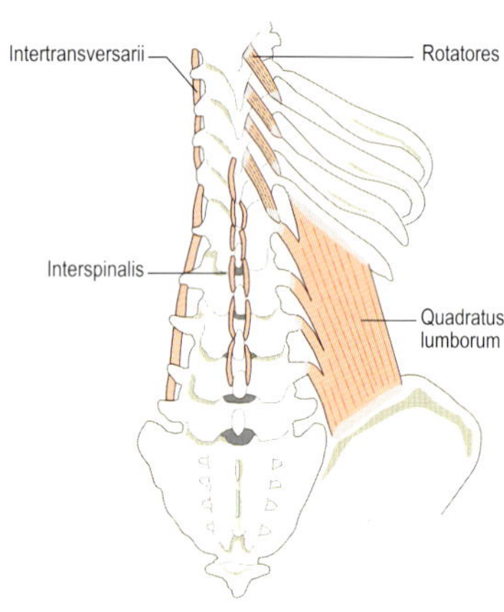

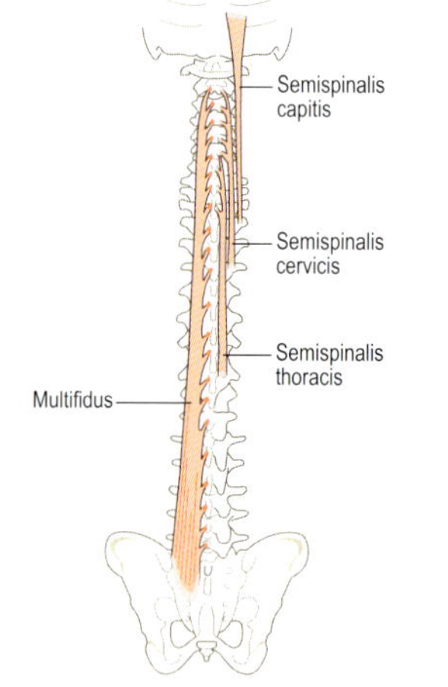

Posterior views

Quadratus lumborum – both sides extend lumbar spine; one will laterally flex to same side or in reversed origin/insertion hold pelvis level when standing on one leg.
Origin: iliolumbar ligament, adjacent iliac crest.
Insertion: medial half 12th rib, transverse processes lumbar vertebrae.
Nerve supply: anterior rami T12–L4.

Multifidus – rotates, extends and laterally flexes spine.
Origin: posterior sacrum, mamillary processes lumbar vertebrae, transverse processes thoracic vertebrae, articular processes lower five cervical vertebrae.
Insertion: muscle fibres arranged in three layers passing upwards/medially to attach to all spinous processes (L5–C1).
Nerve supply: posterior rami adjacent spinal nerves.

Semispinalis – rotates and extends spine. Consists of three parts: *thoracis* – transverse processes T6–10 to spinous processes C6–T2; *cervicis* – transverse processes T1–6 to spinous processes C2–6; *capitis* – transverse processes T1–6 and articular processes C4–7 to nuchal lines on skull.
Nerve supply: posterior rami adjacent spinal nerves.

Intertransversarii and interspinales – intertransversarii laterally flex spine, interspinales extend spine: both are slips of muscle passing between transverse or spinous processes respectively; they also stabilize adjacent vertebrae.

Rotatores – rotate spine and are best developed in thoracic region; run from transverse process to lamina of vertebra above.
Nerve supply: intertransversarii, interspinales and rotatores supplied by posterior rami adjacent spinal nerves.

ABDOMINAL MUSCLES

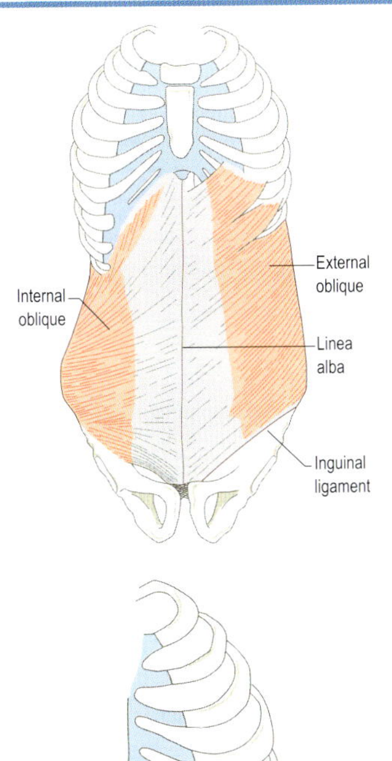

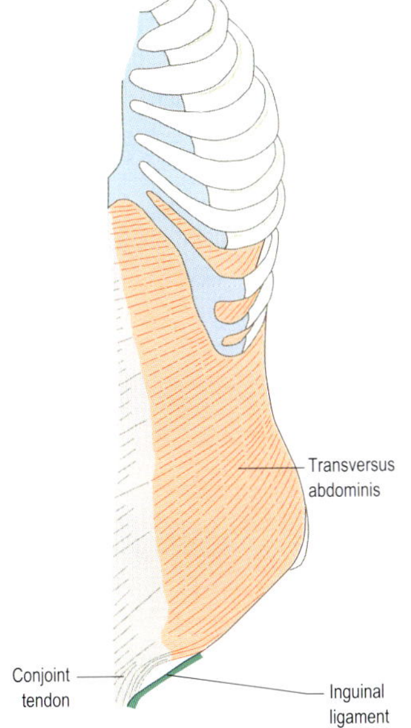

Anterior view oblique and transverse abdominals

External oblique – outermost sheet with fibres running obliquely downwards and medially.
Origin: outer borders lower 8 ribs and their costal cartilages.
Insertion: outer lip anterior two-thirds iliac crest forming large aponeurosis joining that from other side at *linea alba*, a fibrous band running from xiphoid to pubic symphysis; the free lower border forms the *inguinal ligament* which runs between anterior superior iliac spine and pubic tubercle.

Internal oblique – middle layer with fibres running upwards and medially.
Origin: lateral two-thirds inguinal ligament, anterior two-thirds iliac crest and thoracolumbar fascia (formed by fascia covering psoas major, quadratus lumborum and erector spinae).
Insertion: inferior border lower 4 ribs, forms aponeurosis which joins its fellow at *linea alba*; the portion from the inguinal ligament joins with that of transversus abdominis forming a conjoint tendon attaching to pubic crest.

Transversus abdominis – deepest layer with fibres running transversely.
Origin: lateral third inguinal ligament, anterior two-thirds inner lip iliac crest, thoracolumbar fascia, inner surface lower 6 ribs and their costal cartilages.
Insertion: forms aponeurosis attaching to its fellow at linea alba; the portion from the inguinal ligament forms the conjoint tendon with internal oblique.

Nerve supply: all three muscles supplied by lower 6 or 7 thoracic and 1st lumbar nerves.

ABDOMINAL MUSCLES

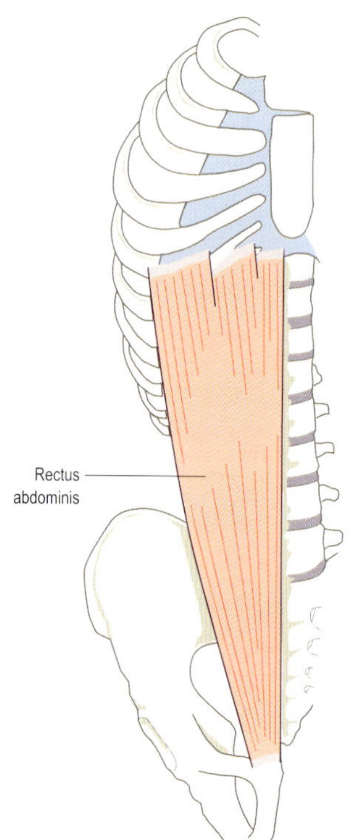

Anterior view, right side

Rectus abdominis – runs down front of abdomen, enclosed in rectus sheath. Three tendinous intersections cross muscle transversely to give the 'six-pack' appearance.

Origin: front of symphysis pubis and pubic crest.

Insertion: xiphoid process, costal cartilages of 5th, 6th, 7th ribs.

Nerve supply: lower 6 or 7 thoracic nerves.

Actions of abdominal muscles

- Flexion of trunk – produced by internal and external obliques and recti of both sides. In 'reversed' action, i.e. ribs fixed, the pelvis is lifted, altering the 'pelvic tilt'.
- Lateral flexion of trunk – produced by external and internal obliques and rectus of same side. In 'reversed' action the pelvis on the same side is lifted (as in walking).
- Rotation of trunk to left is produced by left internal and right external obliques.

Abdominals work together to raise intra-abdominal pressure. With diaphragm fixed this produces the '*expulsive acts*' of vomiting, parturition, defecation and micturition. In coughing and sneezing the pressure increase is caused by the abdominals forcing the diaphragm upwards. They also contribute to trunk stability during a range of activities.

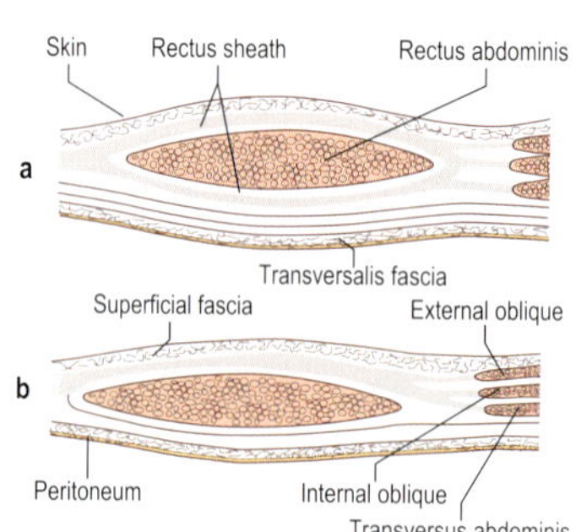

Formation of rectus sheath:
(a) upper and (b) lower abdomen

Rectus sheath – formed from the aponeuroses of abdominal muscles which enclose the two rectus abdominis muscles and join centrally at linea alba. The arrangement of layers changes in the lower abdomen as shown.

During late pregnancy the linea alba is stretched. If it fails to return to normal width the result is called *diastasis recti*.

FASCIA

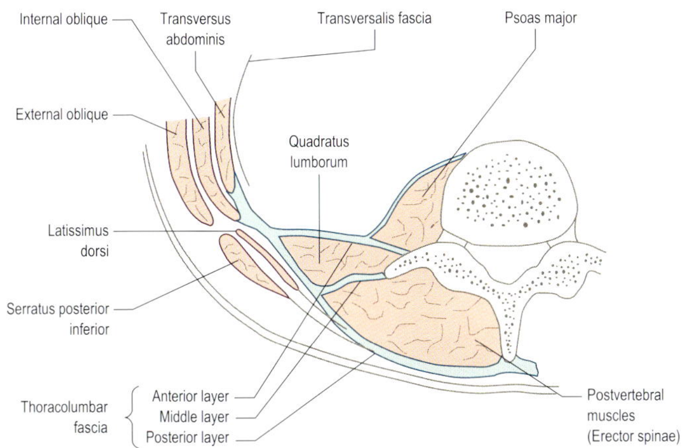

Formation of thoracolumbar fascia

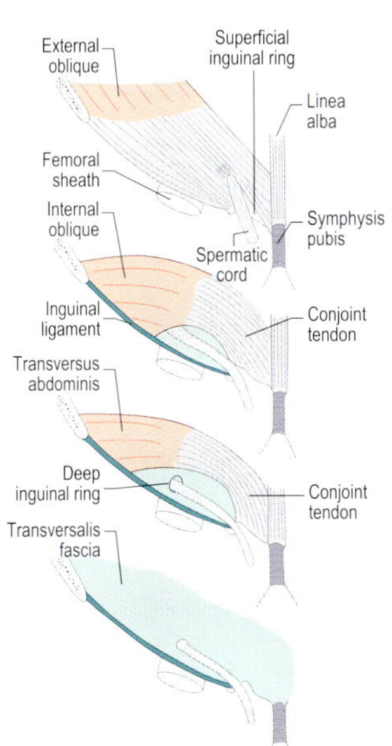

Formation of inguinal canal

The *thoracolumbar fascia* covers psoas major, erector spinae and quadratus lumborum, forming a free edge running between 12th rib and iliac crest. Two abdominal muscles attach to this edge (as shown).

The *inguinal canal* is a 4 cm-long oblique passage through the abdominal wall. It runs from *deep inguinal ring* in transversalis fascia to *superficial inguinal ring* in the aponeurosis of external oblique at medial end of inguinal ligament. The floor of the canal is the inguinal ligament, anterior wall the external oblique aponeurosis, posterior wall is transversalis fascia reinforced by the conjoint tendon and the roof formed by arching fibres of internal oblique and transversus abdominis.

The spermatic cord (male) or round ligament of uterus (female) and ilioinguinal nerve pass through the canal.

The inguinal canal is a weak point in the anterior abdominal wall. Occasionally increased intra-abdominal pressure can force abdominal viscera through it, forming an *inguinal hernia.*

SUMMARY – MUSCLES AND MOVEMENTS

TRUNK

Flexion

Rectus abdominis
External and internal obliques
Psoas major and minor

Extension

Quadratus lumborum
Multifidus
Semispinalis
Erector spinae

Rotation

Internal oblique
External oblique
Multifidus
Rotatores
Semispinalis

Lateral flexion

Quadratus lumborum
Intertransversarii
External oblique
Internal oblique
Rectus abdominis
Multifidus

MUSCLES OF RESPIRATION

Inspiration

Diaphragm
Intercostals
Levatores costorum
Serratus posterior superior

Expiration

Transversus thoracis
Subcostals
Serratus posterior inferior
External and internal obliques
Transversus abdominis
Latissimus dorsi

HEAD AND NECK

Flexion

Longus colli
Sternomastoid
Scalenus anterior
Longus capitis
Rectus capitis (head only)

Lateral flexion

Scalenus anterior, medius and posterior
Levator scapulae
Sternomastoid
Splenius capitis
Trapezius
Erector spinae
Rectus capitis lateralis (head only)

Extension

Levator scapulae
Splenius cervicis
Trapezius
Splenius capitis
Erector spinae
Rectus capitis posterior major and minor (head only)
Superior oblique (head only)

Rotation

Semispinalis cervicis
Multifidus
Scalenus anterior
Splenius cervicis and capitis
Sternomastoid
Inferior oblique (head only)
Rectus capitis posterior major (head only)

NERVE SUPPLY: THE CERVICAL PLEXUS

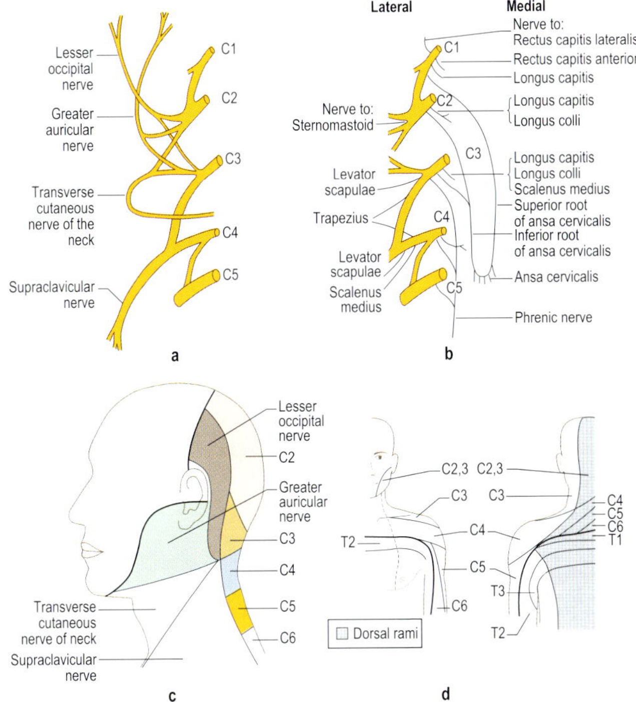

Cervical plexus: (a) superficial (cutaneous); (b) deep (motor) branches; (c) cutaneous distribution; (d) dermatomes

The above diagrams show formation and distribution of the cervical plexus (C1, 2, 3, 4), with C5 being part of the brachial plexus.

Superficial branches are shown in (a) and their cutaneous supply in (c). In (b) deep branches are shown with the names of muscles they supply. Dermatomes are shown in (d).

Phrenic nerve (C3, 4, 5) There is a left and right phrenic nerve passing through the neck, thorax and diaphragm to supply the diaphragm from its inferior surface. The major part of the nerve comes from C4.

> The high root-level supply to the diaphragm means it remains innervated in spinal lesions up to C4/5 segments. It is possible to sever one phrenic nerve, leaving half the diaphragm working.

NERVE SUPPLY: INTERCOSTAL NERVES

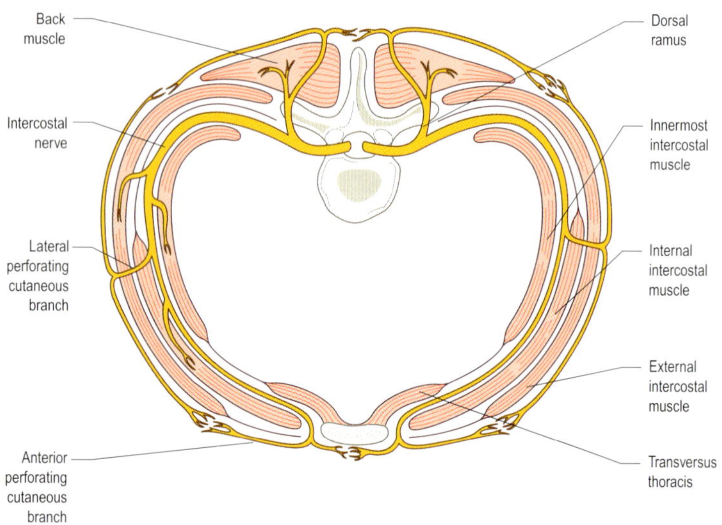

Course and distribution of a typical intercostal nerve

Typical intercostal nerves pass laterally from the intervertebral foramen below the rib of the same segment. Each innervates the intercostal muscles in the space; the lower six extend into the anterior abdominal wall and innervate the four abdominal muscles on each side. The sensory supply of these nerves is represented as a series of transverse parallel bands around the trunk.

The posterior rami (dorsal rami) pass backwards to supply the skin and muscles of the back.

NERVE SUPPLY: SPINAL SEGMENTS AND SPINAL NERVES

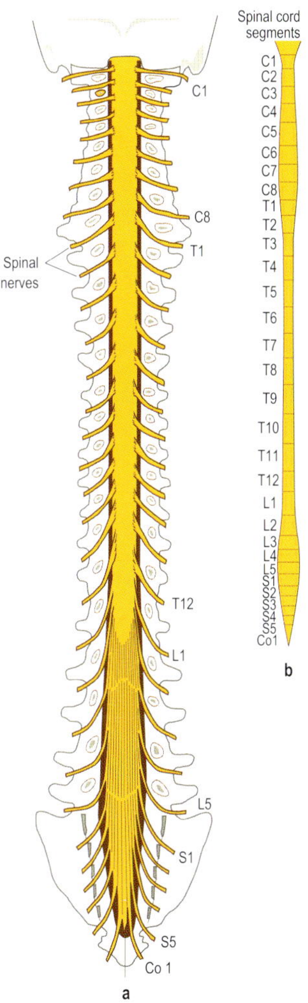

(a) The spinal cord and spinal nerves viewed from behind with laminae removed; (b) the spinal cord divided into its segments

In early development the spinal cord is the same length as the bony vertebral column with each spinal nerve running horizontally. As the foetus develops, differential growth results in a longer bony column than spinal cord. The dural sac grows to accommodate this, terminating at S2, whilst the spinal cord ends at the level of L1/2 disc.

Although a continuous structure, it is convenient to divide the spinal cord into *segments*, each receiving sensory and giving off motor nerve roots on either side.

In the adult, nerve roots still leave via their appropriate intervertebral foramina. In the cervical spine the upper roots are horizontal, but in the lumbar and sacral regions the nerve roots descend some distance from their cord segment to their intervertebral foramina. Below L2 only descending nerve roots are found in the vertebral canal, forming the so-called *cauda equina* ('horse's tail').

Diagnosis of level of spinal cord injury is complicated by the differential relationship between bony and neurological levels.

Trauma/pathology of motor cells in the cerebral cortex produce the effects of an upper motor neuron lesion, while trauma/pathology of motor cells in the spinal cord produce the effects of a lower motor neuron lesion.

NERVE SUPPLY: SPINAL NERVES AND PLEXUSES

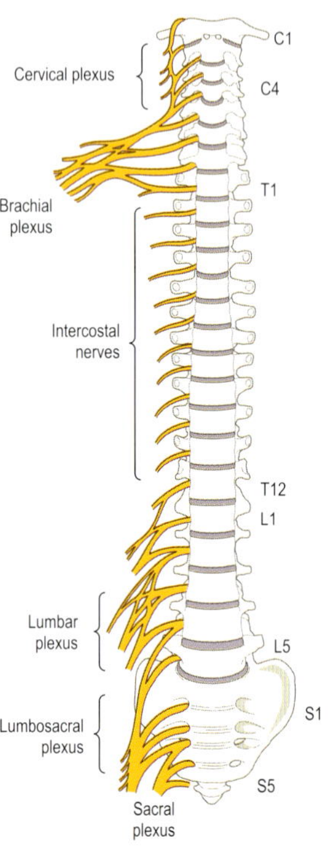

A *plexus* is the joining together of anterior (ventral) rami to form peripheral nerves.

In the thoracic region the *intercostal nerves* run singly between ribs and do not form a plexus.

The *cervical plexus* is formed by C1–4; the *brachial plexus* by C5–T1; the *lumbar plexus* by L1–4; the *lumbosacral plexus* by L4–S3 and the *sacral plexus* by S3–5.

Details of these plexuses are given elsewhere in this book.

Ventral and *dorsal roots* attach to the spinal cord in the vertebral canal and join to form a *spinal nerve* which exits via the intervertebral foramen. The spinal nerve then divides into *anterior (ventral)* and *posterior (dorsal) rami.*

The posterior rami supply skin and muscles of back and neck, whereas the anterior rami mainly form plexuses to supply muscles and skin of the limbs and trunk.

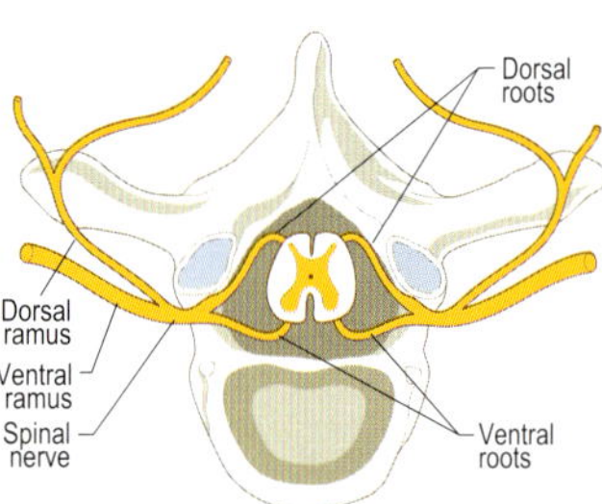

(a) Formation of the major nerve plexuses by anterior primary rami of spinal nerves; (b) formation of a spinal nerve and ventral and dorsal rami

NERVE SUPPLY: INTERNAL STRUCTURE AND ARTERIAL SUPPLY OF SPINAL CORD

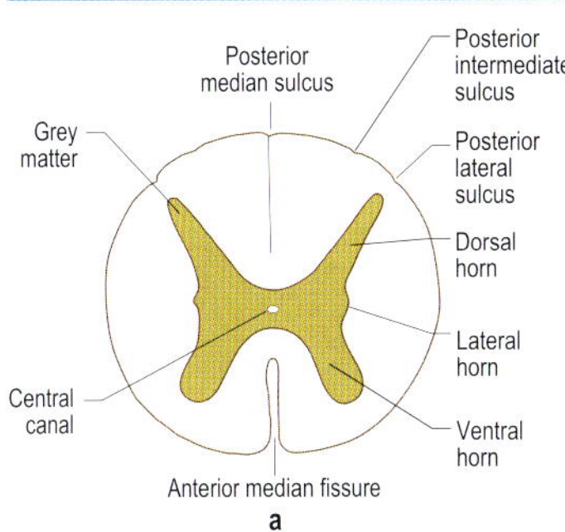

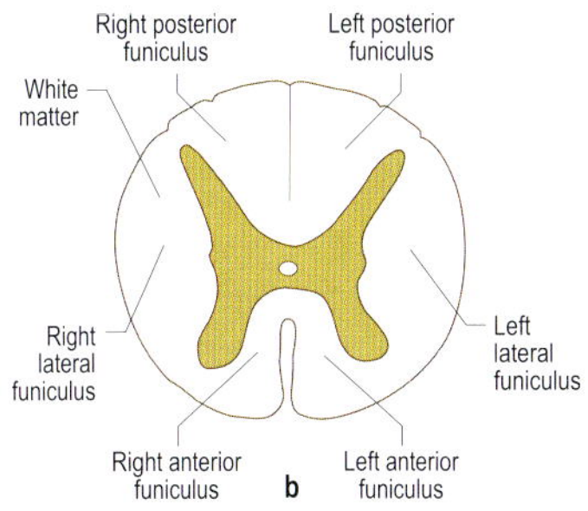

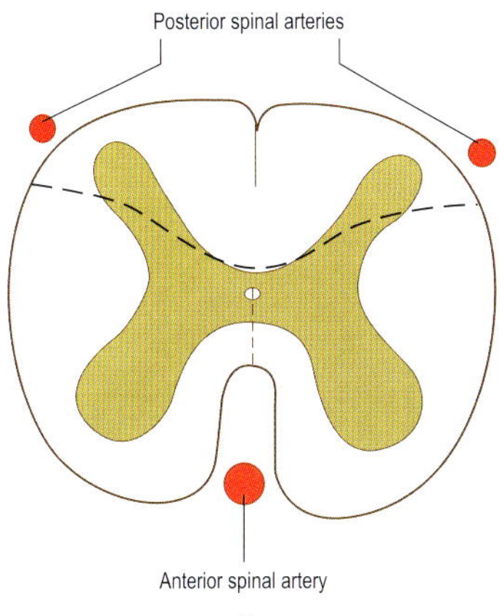

The central core of the spinal cord consists of *grey matter* made up of the cell bodies of neurons. The *ventral (anterior) horns* contain cells that innervate voluntary muscle and the axons from these leave via the ventral roots. The *dorsal (posterior) horns* contain cells concerned with sensory information that enters through the dorsal roots. Between the T1 and L2 segments a small *lateral horn* is formed by cell bodies of neurons forming part of the autonomic nervous system.

The peripheral part of the spinal cord consists of *white matter* formed by axons arranged in tracts that convey information to or from the brainstem or between segments of the spinal cord. The white matter is divided into three regions: a *lateral*, *posterior* and *anterior funiculus.*

The arterial supply of the spinal cord is derived from the vertebral arteries which give branches to form a single *anterior spinal artery* which descends along the front of the cord. Most of the blood supply is derived from this vessel but smaller branches arise from intercostal and lumbar arteries at variable levels along the cord.

The smaller *posterior spinal arteries* pass along the posterolateral aspect of the cord supplying a small adjacent area.

Transverse sections of the spinal cord to show (a) grey matter and sulci (labelled), (b) white matter (labelled) and (c) arterial supply

NERVE SUPPLY: THE SPINAL MENINGES

The spinal cord is surrounded and protected by the meninges, and is bathed in cerebrospinal fluid (CSF). The outer *dura mater* is tough and fibrous, attaching to the margins of the foramen magnum and stretching to S2.

The deepest layer, *pia mater*, is thin and wraps like a skin around the spinal cord. The middle layer, *arachnoid mater*, lines the dura leaving a space (*subarachnoid space*) between it and the pia mater; this space is permeated by a network of threads ('spider's web'). In this space the pia forms 21 pairs of *denticulate ligaments* and the *filum terminale* which anchor the spinal cord inferiorly.

The subarachnoid space is filled with CSF which is continuously secreted from the brain's ventricles, entering the subarachnoid space through the 4th ventricle. It circulates around the brain and spinal cord and is reabsorbed through the arachnoid granulations. This fluid protects the brain and spinal cord.

The emerging spinal nerve roots in penetrating the meninges carry the dura mater with them for a short distance as the *dural sleeves*.

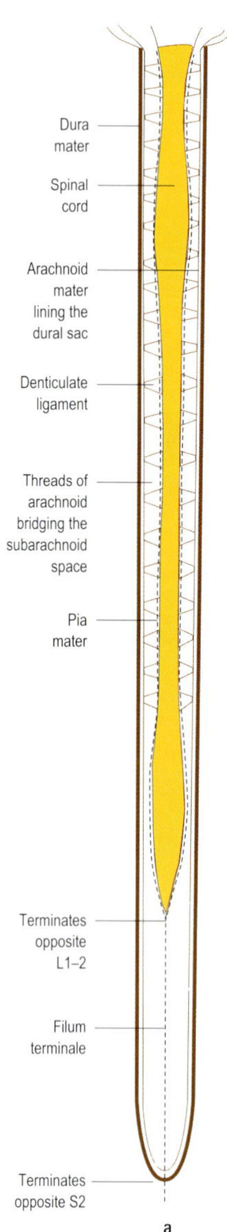

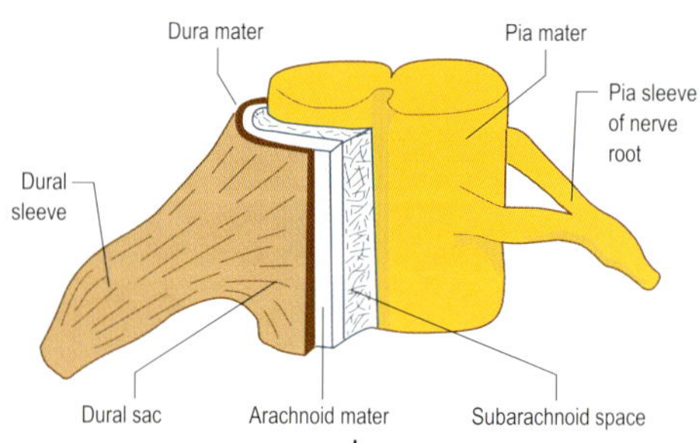

The meninges showing (a) the arrangement around the spinal cord and (b) in section, around a spinal nerve

NERVE SUPPLY: AUTONOMIC NERVOUS SYSTEM, SYMPATHETIC CHAIN AND EFFECTS

Outflow for the sympathetic nervous system is from thoracolumbar (T1–L2) parts of the spinal cord. The main neurotransmitter is noradrenaline.

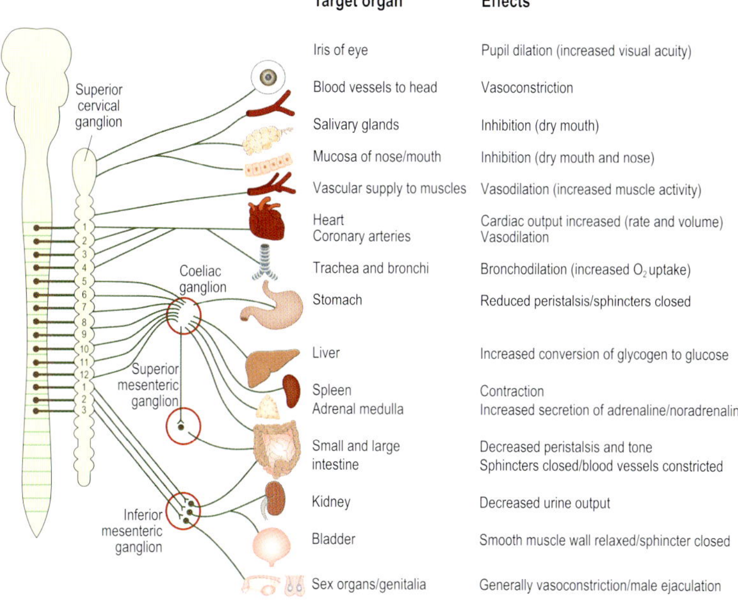

NERVE SUPPLY: AUTONOMIC NERVOUS SYSTEM, PARASYMPATHETIC OUTFLOW AND EFFECTS

Outflow for the parasympathetic nervous system is from cranial nerves (III, VII, IX, X) in the brainstem and from spinal nerves S2, 3, 4 (craniosacral outflow). The main neurotransmitter is acetylcholine.

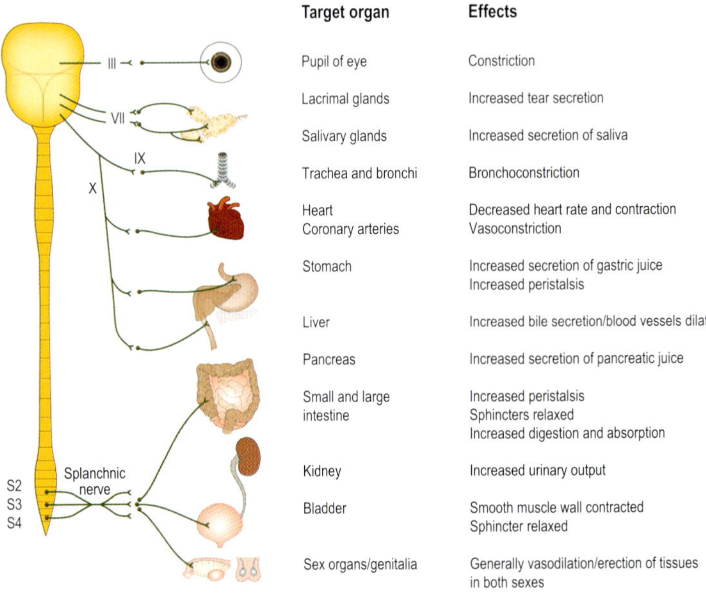

ARTERIAL SUPPLY

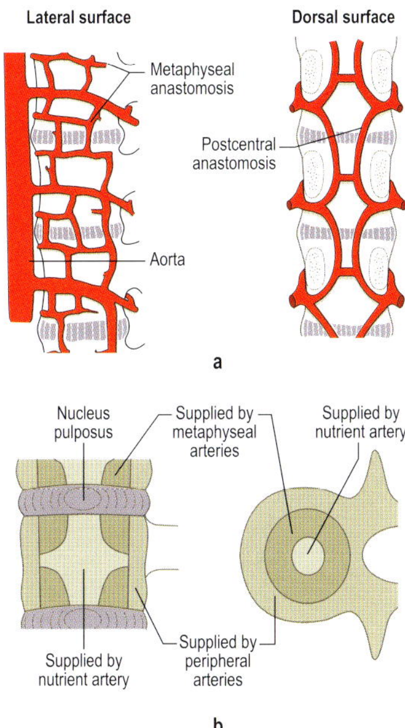

(a) Anastomoses between lumbar arteries and their branches; (b) zoning of blood supply to the intervertebral disc. (Adapted from Radcliffe JF (1980) The arterial anatomy of the adult human lumbar vertebral body: a microarteriographic study. *Journal of Anatomy*, 131, 57–79.)

The vertebral column has a segmental arterial supply via branches of vessels adjacent to it. In the cervical region these are from vertebral and ascending cervical arteries in the thoracic region from posterior intercostal arteries in the lumbar region from lumbar and iliolumbar arteries and in the pelvis from lateral sacral arteries. All of these anastomose with the anterior and posterior spinal arteries supplying the spinal cord.

VENOUS DRAINAGE

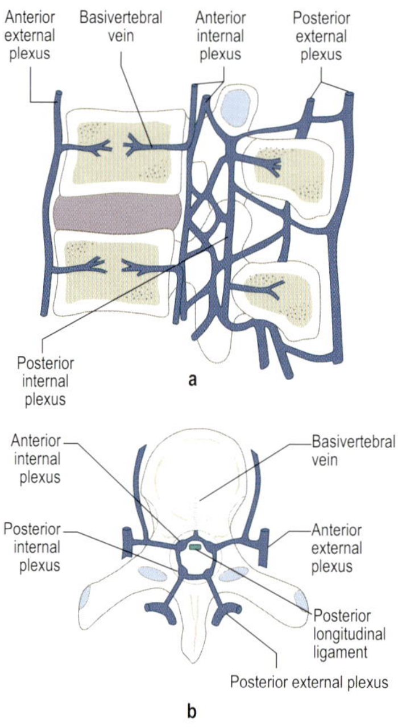

Venous drainage of the vertebral column: (a) sagittal section; (b) transverse section

The veins of the vertebral column form complex communicating plexuses extending the whole length of the column both inside and out.

These plexuses drain via intervertebral veins into the vertebral veins of the neck, intercostal veins in the thorax, lumbar veins in the back and sacral veins in the pelvis.

ABDOMEN AND PELVIS

INTRODUCTION

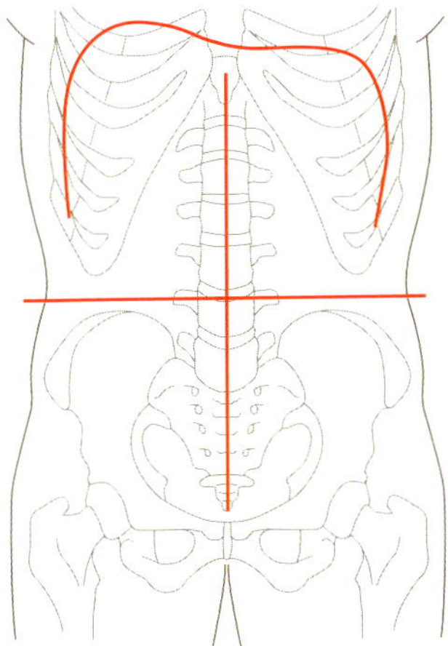

Abdominal quadrants

Right upper	Left upper
Liver Gall bladder Colon (hepatic flexure and transverse) Kidney and adrenal gland Duodenum with head of pancreas Small intestine	Stomach Spleen Pancreas Kidney and adrenal gland Colon (splenic flexure and transverse) Small intestine (jejunum)
Right lower	**Left lower**
Colon (ascending) Caecum Appendix Small intestine	Colon (descending) Sigmoid colon Small intestine

The four abdominal quadrants

The abdomen and pelvis contain the major parts of the digestive and urogenital systems, together with major blood vessels and nerves: it is separated from the thorax by the diaphragm and the perineum by levator ani.

The abdomen can be divided into quadrants as shown above, centred on the umbilicus. Although the position of the umbilicus is variable, the abdominal quadrants are extensively used in clinical practice.

An alternative way to divide the abdominal regions is to use two vertical midclavicular lines and horizontal transpyloric and transtubercular planes, giving nine abdominal regions.

DIGESTIVE SYSTEM

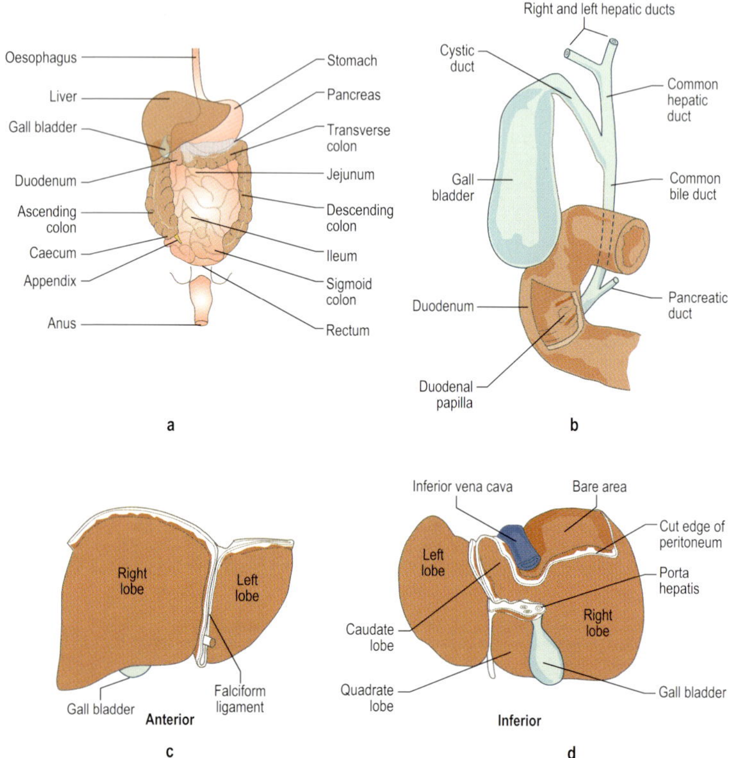

(a) Organs of the digestive system within the abdomen; (b) the biliary system; (c) the liver viewed anteriorly and (d) inferiorly

The digestive tract is effectively a long tube extending from the oral cavity to the anus and comprises the mouth, pharynx, oesophagus, stomach, small and large intestines. It has a moist epithelial lining containing numerous glands surrounded by inner circular and outer longitudinal muscle layers, which move the ingested food and secreted enzymes along the tract (peristalsis).

The liver is a large solid organ, part of whose function is to produce bile, which is stored in the gall bladder and released into the duodenum via the common bile duct. The pancreas is both an exocrine and an endocrine gland, the exocrine secretion also entering the duodenum via the pancreatic duct. Both organs contribute to the digestive process.

HERNIAE

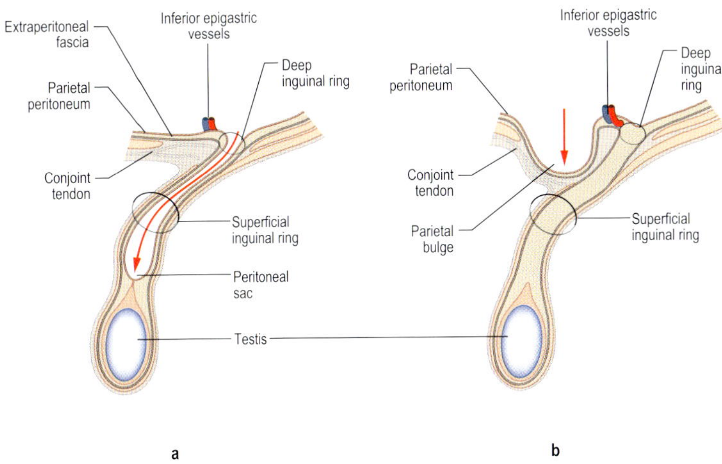

(a) Indirect and (b) direct inguinal herniae

A hernia is a protrusion of part of the digestive tract into a space where it is not normally found: it may be internal or external. *Hiatus hernia* is the commonest internal hernia in which either the gastro-oesophageal junction and upper stomach or part of the stomach pass through the diaphragm.

The commonest external hernia is an *inguinal hernia* (75%), then *femoral hernia*. Most inguinal hernias are indirect (predominantly in males) in which a loop of small intestine passes along the inguinal canal through the anterior abdominal wall. Once through the superficial inguinal ring the hernia enters the scrotum/labia major. In direct inguinal hernias the protruding gut pushes through the posterior wall of the inguinal canal.

Femoral hernias are more common in females. A loop of gut pushes through the femoral ring into the femoral sheath, appearing below the inguinal ligament lateral to the pubic tubercle: the hernia is limited by the fascia lata except at the saphenous opening through which it may pass. With all external herniae there is a danger of strangulation, in which the intestine becomes twisted and its blood supply cut off.

URINARY SYSTEM

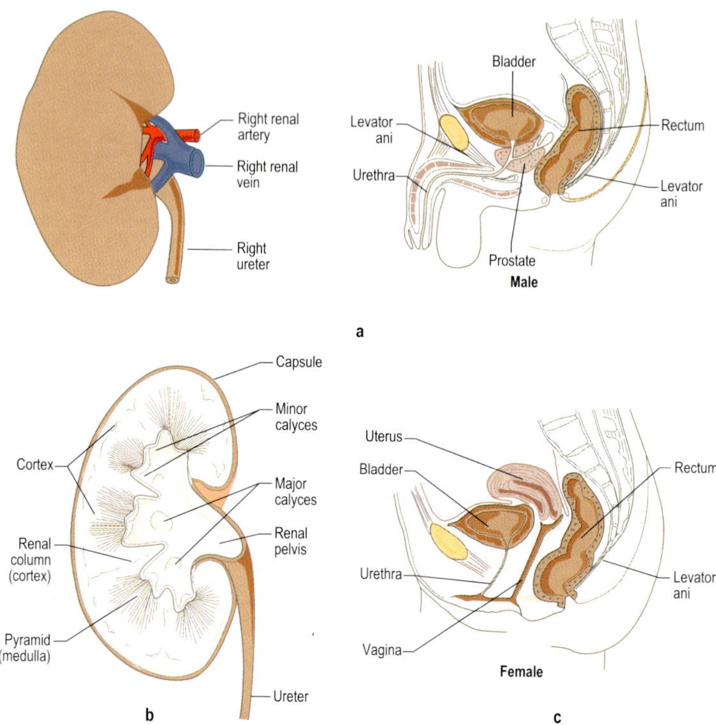

Right kidney: (a) anterior view and (b) internal organization; (c, d) relations of the bladder in male and female

The urinary system lies partly in the abdomen (kidneys, ureters), partly in the pelvis (bladder), with the remainder passing through the perineum (urethra). The kidneys lie on the posterior abdominal wall in paravertebral gutters alongside the upper three lumbar vertebrae with the diaphragm superiorly. The ureter passes inferiorly over psoas and the pelvic brim anterior to the bifurcation of the common iliac artery entering the bladder obliquely at the upper lateral angle of the trigone. Along its length the ureter has three constrictions (junction with renal pelvis, as it crosses the pelvic brim, as it enters the bladder) where kidney stones may lodge. The bladder is a hollow organ lying behind the symphysis pubis: in infants it is an abdominal organ. The urethra leaves the bladder at the inferior angle of the trigone. In males it passes through the prostate (prostatic urethra), perineal membrane (membranous urethra) and penis (penile or spongy urethra): in females it passes through the pelvic floor and perineal membrane, being firmly attached to the anterior vaginal wall.

FEMALE REPRODUCTIVE SYSTEM

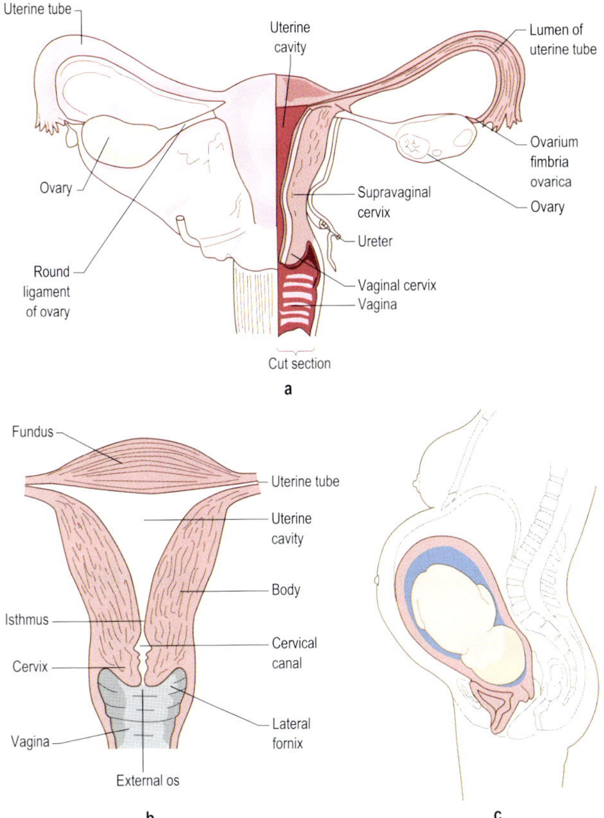

(a) The uterus and ovaries (broad ligament removed); (b) the uterus (coronal section); (c) the enlarged uterus near full term

The ovaries, uterine tubes, uterus and upper part of the vagina of the female reproductive system lie in the pelvis, with the lower part of the vagina being in the perineum.

During pregnancy the foetus and uterus occupy more and more of the abdominal cavity. From being a pelvic organ the uterus reaches the symphysis pubis by week 12, the umbilicus by week 24 and xiphisternum by week 36. In the last 4 weeks the foetal head may descend into the pelvis (engagement of the head) and the fundus may also descend. After birth the uterus normally returns to its former size within 6 weeks.

THE PELVIC FLOOR

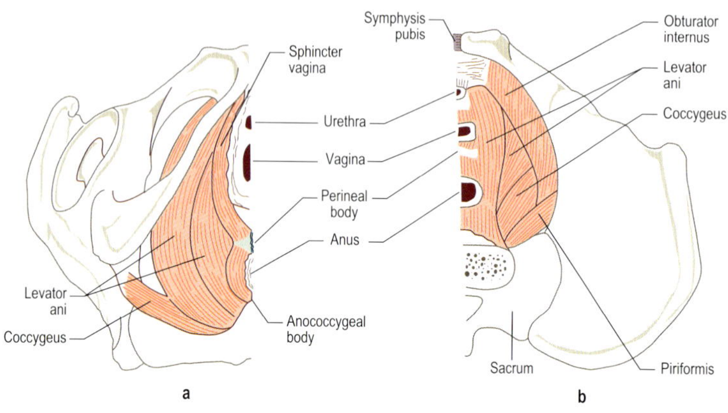

Female pelvic floor: (a) inferior and (b) superior views

Levator ani

Origin: each muscle arises in continuous manner from inner surface of pubic body and obturator membrane, pelvic surface of ischial spine.
Insertion: considered in two parts: *iliococcygeus* and *pubococcygeus* which sweep towards midline inserting into perineal body, anal canal, anococcygeal body and sides of coccyx; anterior fibres pass around prostate (male), vagina (female) and anus.
Nerve supply: pudendal nerve S4, anterior rami S3, 4.

Coccygeus – lies posterior to, and in same plane as, levator ani.
Origin: ischial spine.
Insertion: margin of coccyx and lower two sacral segments.
Nerve supply: anterior rami S4. Both muscles support pelvic viscera and are constantly active. Coccygeus pulls coccyx forwards following defecation.

Contraction of levator ani constricts openings in the pelvic floor, either reflexly to resist increased pressure (coughing) or voluntarily when assisting sphincters to prevent inconvenient micturition or defecation. In the female, levator ani supports the vagina and uterus.

The pelvic floor is stretched during childbirth and muscles may be torn or surgically cut (episiotomy) to enlarge the birth canal. Following birth, the pelvic floor must be exercised to regain normal tone and function; *stress incontinence* may result if this is not achieved. Following prostate surgery similar problems may occur in the male.

PART 5

The head and brain

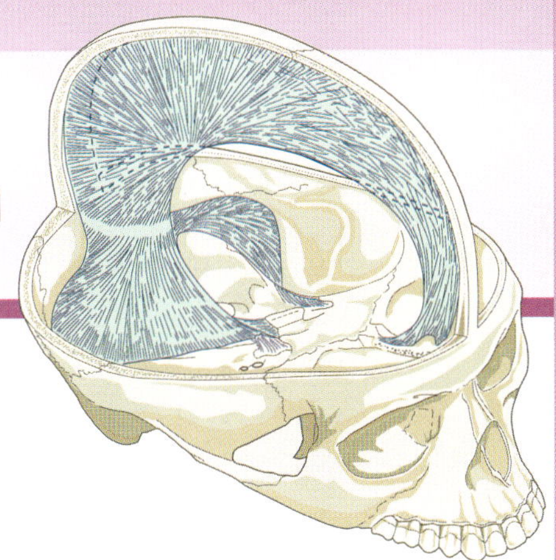

SKULL

INTRODUCTION

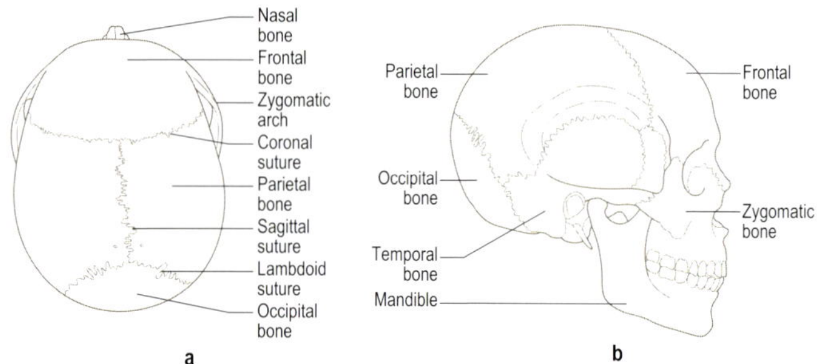

(a) Superior and (b) lateral views of adult skull

The skull is a complex arrangement of bones forming the head and face. The large cranial cavity encloses the brain and the facial skeleton forms the walls of the orbits, nasal cavity and roof of the mouth: the mandible is a separate bone.

At birth individual bones are joined by fibrous tissue: the sagittal and coronal sutures meet at the anterior fontanelle which closes by 18 months; the sagittal and lambdoid sutures meet at the posterior fontanelle which closes by 3 months; the parietal, frontal, temporal and greater wing of the sphenoid meet laterally at the pterion, which also closes by 3 months.

After birth the skull grows rapidly until the 7th year, after which growth slows down.

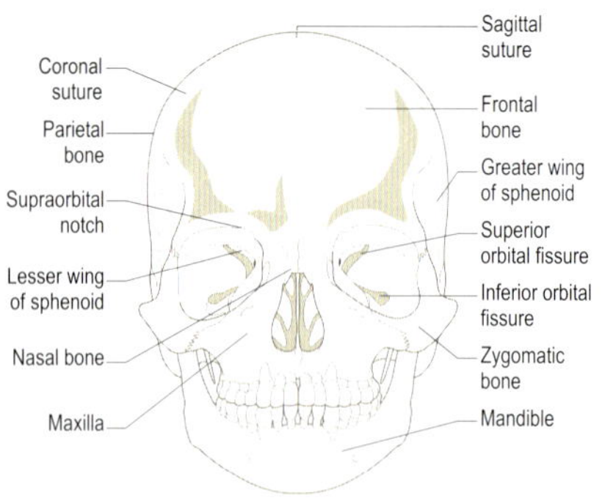

Anterior view adult skull

Palpation

Most of the skull can be easily palpated. Posteriorly the external occipital protuberance with the superior nuchal lines running laterally can be identified. Laterally the external acoustic meatus is visible, behind which is the mastoid process. Above the meatus the posterior part of the zygomatic arch can be identified passing forward to join the zygomatic bone. The margins of the orbit can also be palpated.

MANDIBLE

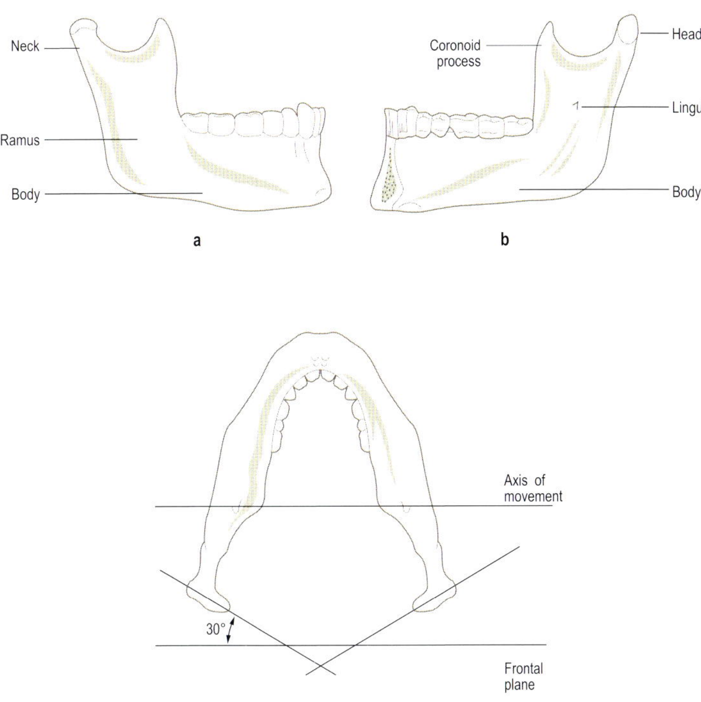

(a) Lateral, (b) medial and (c) inferior views of the mandible

The horizontal convex body and upward-projecting rami of the mandible have lateral and medial surfaces. The body has superior and inferior borders, the rami anterior and posterior borders: where the body meets the ramus is the angle. On the medial surface of the body is the mylohyoid line, with the mental spines anteriorly: the superior border houses the teeth. The concave superior border passes between the posterior condylar (head) and anterior coronoid processes. The long axes of each head converge posteromedially.

The entire length and depth of the mandible are palpable, with the posterior angle being prominent, especially in males. The line of the temporomandibular joint can be identified by placing the tip of the finger immediately anterior to the tragus of the ear. As the mouth is opened the condyle moves forwards and a large depression can then be felt – this is the joint cavity.

TEMPOROMANDIBULAR JOINT

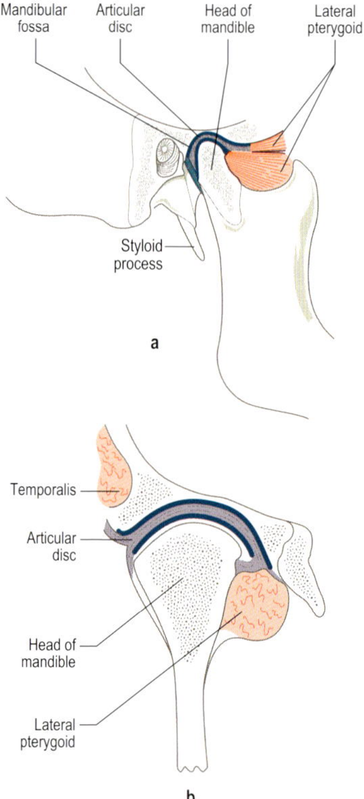

Temporomandibular joint: (a) sagittal and (b) coronal sections

Synovial condyloid joint between the head of mandible and the mandibular fossa of the temporal bone on the base of the skull: both articular surfaces are covered by fibrocartilage. The spindle-shaped head lies at right angles to the ramus: it is smaller than the mandibular fossa. The oval temporal fossa is concavoconvex from behind forwards, being wider mediolaterally than anteroposteriorly.

An intra-articular disc completely separates the incongruent joint surfaces, moulding itself upon them during movements at the joint.

TEMPOROMANDIBULAR JOINT CAPSULE AND LIGAMENTS

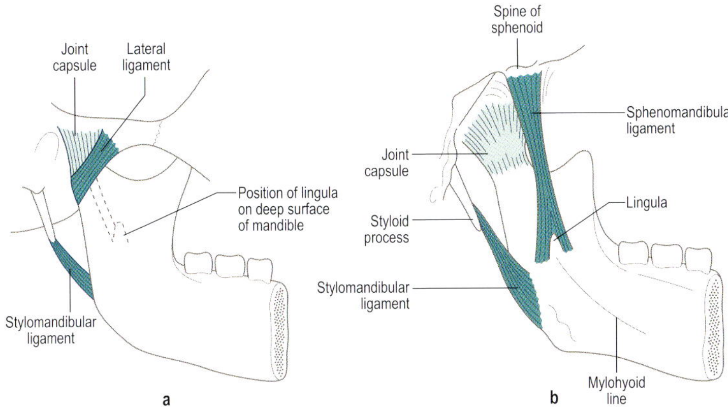

Capsule and ligaments of the temporomandibular joint: (a) lateral and (b) medial views

A strong fibrous capsule attaches to the margins of the mandibular fossa and neck of mandible: the intra-articular disc attaches to its inner aspect. Part of the tendon of lateral pterygoid inserts into the front of the capsule and thus indirectly to the disc. The capsular attachments allow a rotatory movement between the condyle and undersurface of the disc, as well as enabling the disc and condyle to move backwards and forwards together against the mandibular fossa.

Three ligaments are associated with the joint:

Lateral ligament From the lower border and tubercle of the zygomatic arch it passes posteroinferiorly to the lateral and posterior aspect of the mandibular neck, blending with and reinforcing the joint capsule.

Sphenomandibular ligament Strong, thin flat band lying medial to the joint. From the spine of the sphenoid it passes anteroinferiorly to the lingula of the mandible.

Stylomandibular ligament Passes from the apex of the styloid process to the lower part of the posterior border of the ramus of the mandible near the angle.

The sphenomandibular and stylomandibular ligaments guide movement of the mandible against the base of the skull.

MANDIBULAR MOVEMENTS

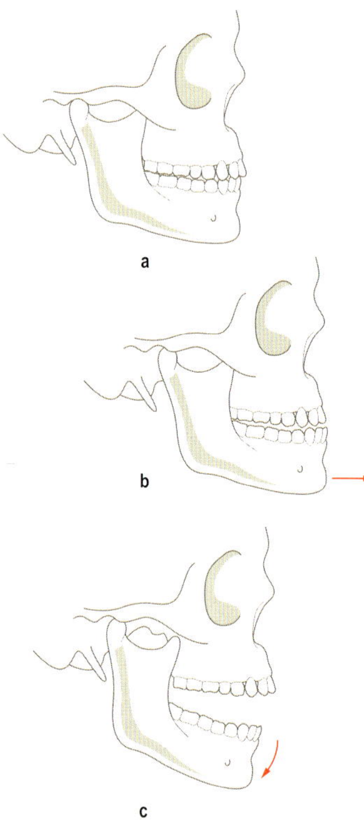

Movement of the mandible: (a) retracted and elevated, i.e. closed position with teeth occluded; (b) protraction; (c) depression

Movements of the mandible are: *depression* (lowering), *elevation* (raising), *protraction* (forward movement), *retraction* (backward movement). All movements are used in chewing.

The presence of an articular disc and shape of the joint surfaces permit different movements in the upper and lower parts of the joint. In the lower compartment a hinge movement occurs between the head of mandible and inferior surface of the disc, while in the upper compartment the head and disc glide forwards and slightly downwards against the mandibular fossa and articular eminence.

In chewing and grinding movements the mandible is alternately protracted and retracted, with the two sides moving in opposite directions. These movements are combined with elevation and depression, resulting in the mandibular teeth moving diagonally across the maxillary teeth with the intervening food being crushed and ground.

Accessory movements: with the subject supine and head turned through 90° a downward pressure applied to the condyle by the thumb moves it transversely within the mandibular fossa. With the head in the same position a forward-directed pressure applied behind the earlobe to the back of the condyle produces forward movement of the condyle.

MUSCLES OF MASTICATION

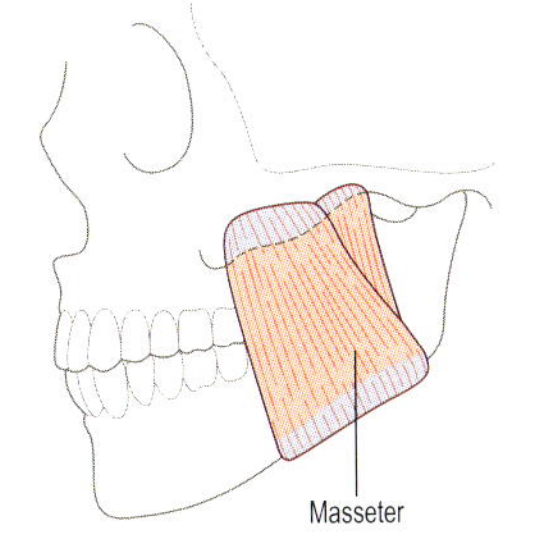

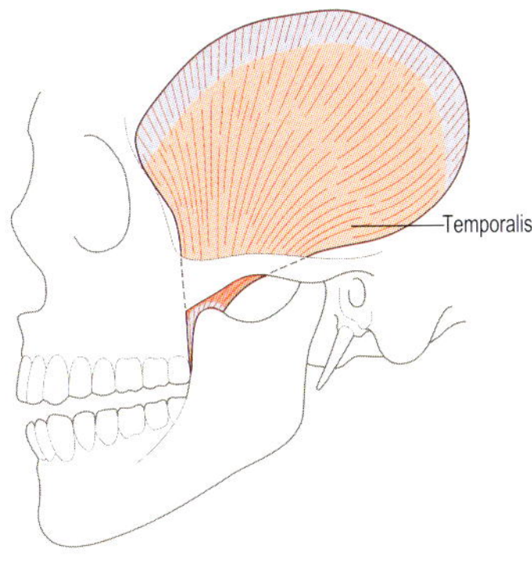

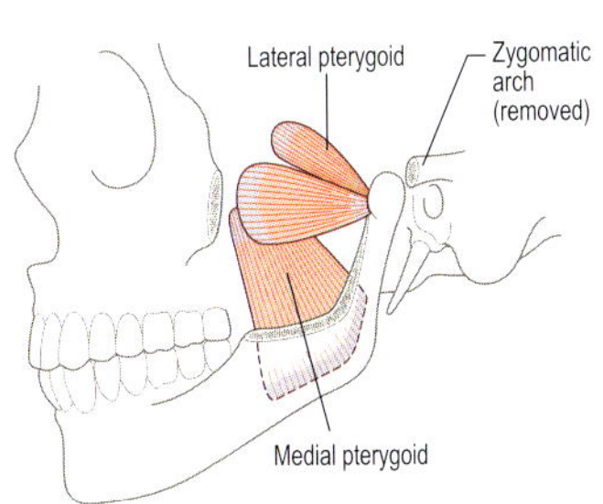

Masseter – elevates and protracts the mandible.
Origin: zygomatic process of maxilla, anterior two-thirds zygomatic arch.
Insertion: outer surface angle of mandible.

Temporalis – elevates and retracts mandible.
Origin: temporal fossa of temporal bone.
Insertion: deep surface of coronoid process and ramus of mandible.

Lateral pterygoid – both sides protract mandible; working with medial pterygoid of same side produces rotation.
Origin: by two heads from inferior surface greater wing of sphenoid and outer surface lateral pterygoid plate.
Insertion: front of neck of mandible.

Medial pterygoid – elevates and protracts mandible; working with lateral pterygoid of same side produces rotation.
Origin: medial side lateral pterygoid plate, palatine bone and maxillary tubercle.
Insertion: inner surface angle of mandible.

Nerve supply: all four muscles supplied by mandibular division of trigeminal (cranial V) nerve.

Functionally these muscles raise the mandible allowing the teeth to bite; they also protract, retract and rotate the mandible to grind food. Some of these muscles oppose gravity and hold the mouth closed.

MUSCLES OF FACE

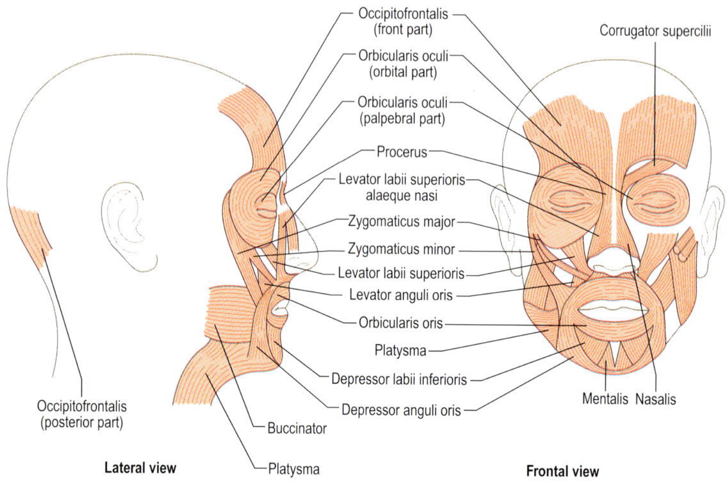

The muscles illustrated above are important for facial expression and moving cheeks and lips during chewing and speech. Some muscles have attachments to bone and skin whereas others attach only to skin and fascia.

The names of these muscles often relate to the part of the face over which they function:

- *occipito* – forehead
- *oculi* – eye
- *nasi* – nose
- *labii* – lips
- *oris* – mouth.

Names may also indicate direction of muscle fibres and therefore pull:

- *orbicularis* – circular/surrounding (as in mouth and eye)
- *levator* – elevates
- *depressor* – depresses.

Muscles not following this nomenclature include:

- *buccinator* – in cheek
- *zygomaticus major* and *minor* – elevate upper lip
- *platysma* – in neck
- *mentalis* and *risorius* – attach to lower lip
- *corrugator* – in forehead.

Nerve supply: all these muscles are supplied by the facial (cranial VII) nerve.

> *Bell's (facial) palsy* results when muscles on one side of the face are paralysed. This causes inability to close the eye fully, accumulation of food between the cheek and lips, drooping of the corner of the mouth causing dribbling; on smiling, the mouth is pulled towards the unaffected side. It also affects speech.

HYOID MUSCLES

The hyoid bone is suspended between two muscle groups, the suprahyoid above and the infrahyoid below.

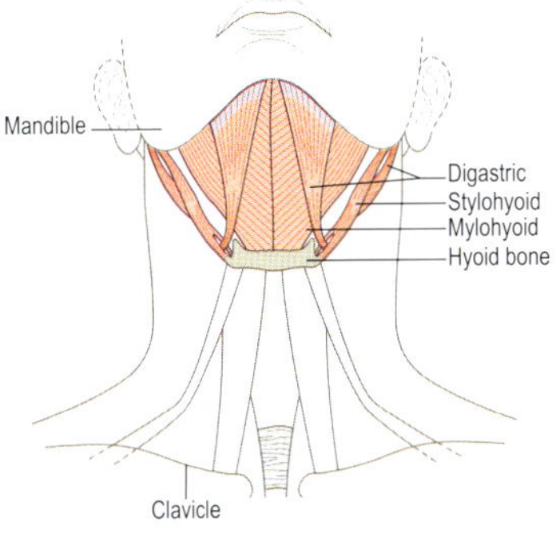

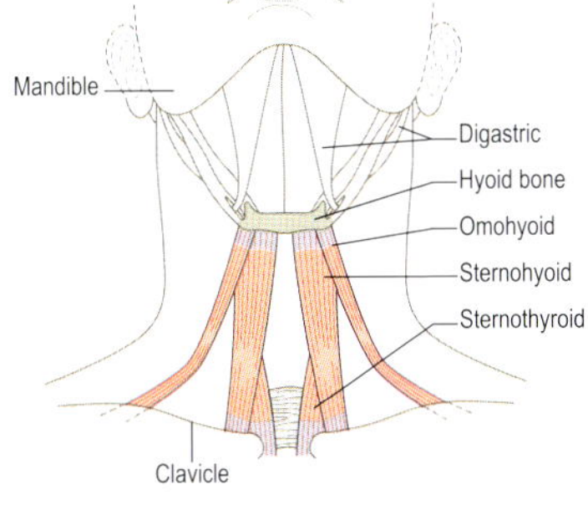

Suprahyoid muscles

Digastric – sling-like with two bellies and central tendinous section. Depresses mandible and elevates hyoid. Bellies attach to mastoid process and lower border of mandible, central part passes through stylohyoid, through which pull is exerted.

Nerve supply: cranial nerves V and VII.
Mylohyoid – depresses mandible and elevates hyoid and floor of mouth. Attaches to inner surface body of mandible, fibres running downwards and medially to insert into its fellow at midline and hyoid (forming floor of mouth).
Nerve supply: cranial nerve V.

Stylohyoid – elevates and retracts hyoid; from styloid process of temporal bone to hyoid.
Nerve supply: cranial nerve VII.

> Elevation of the hyoid is part of swallowing where the larynx is elevated, closing the laryngeal inlet to prevent food passing into the trachea.

Infrahyoid muscles

These pull the hyoid downwards after elevation during swallowing.

Sternohyoid – from medial end clavicle and posterior part manubrium sterni to hyoid.

Sternothyroid – from posterior surface manubrium sterni and 1st costal cartilage to thyroid cartilage (below hyoid).

Thyrohyoid – from thyroid cartilage (as continuation of sternothyroid) to hyoid.

Omohyoid – consists of two bellies united by intermediate tendon (similar to digastric); from suprascapular notch of scapula to hyoid.
Nerve supply: all above muscles supplied by anterior rami C1, 2, 3.

CEREBRAL HEMISPHERES AND CEREBELLUM

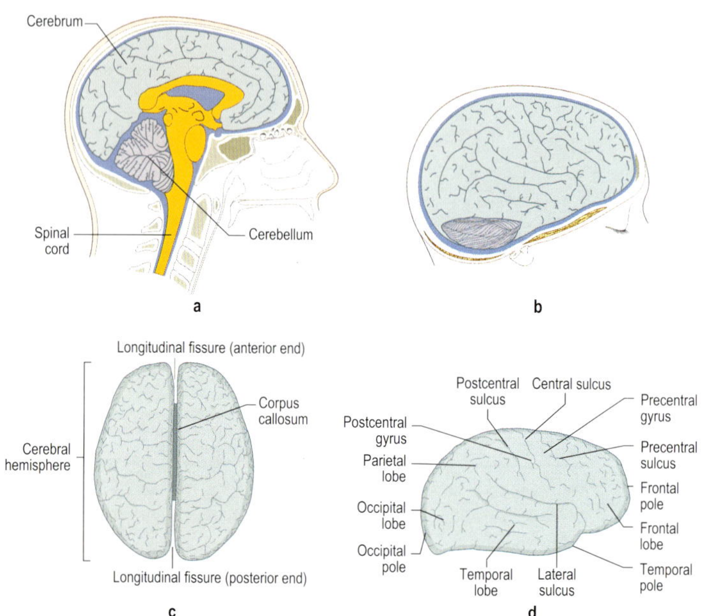

(a) Sagittal section showing the brain within the skull and vertebral canal; (b) brain within the skull; (c) superior and (d) lateral views

The brain lies within the skull: the outer part is formed by nerve fibres and cell bodies (grey matter), the inner part mainly by myelinated axons (white matter). Dispersed between nerve cells are *glial cells* (oligodendrocytes, microglia, ependymal cells, astrocytes) which support the nerve cells: glial cells outnumber nerve cells.

The *cerebral hemispheres* are separated by the longitudinal fissure and joined by the corpus callosum: the surfaces are thrown into ridges (*gyri*) separated by depressions (*sulci*). The precentral gyrus (motor cortex) is responsible for voluntary muscle action, while the postcentral gyrus (sensory cortex) receives sensory information from the brainstem and thalamus. The *occipital lobe* is involved in vision; the *parietal lobe* is responsible for sensory functions (touch, pressure), position of body and limbs, three-dimensional perception, analysis of visual images, language, geometry and calculations. The *frontal lobe* subserves motor functions and governs intellect and personality; the upper part of the *temporal lobe* is involved in the perception of sound, with the remainder concerned with memory and various emotional functions.

The *cerebellum* controls posture, repetitive movements and the accuracy of voluntary movements.

MENINGES

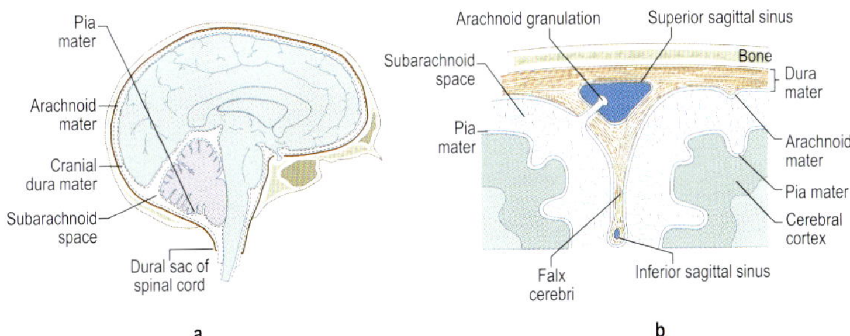

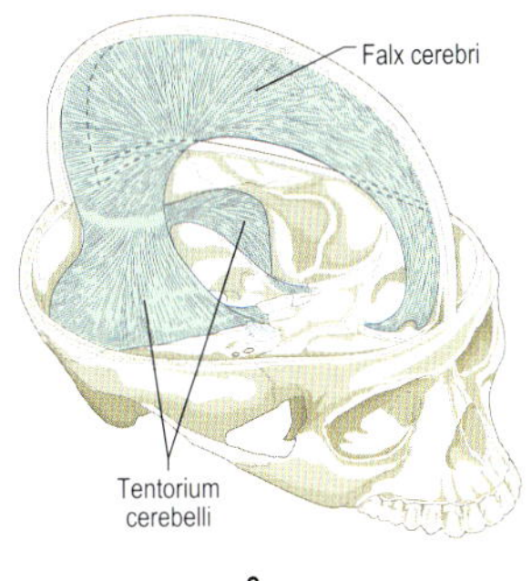

Arrangement of meninges: (a, c) within the skull; (b) in the region of falx cerebri

The brain is surrounded and protected by the meninges. The smooth *pia mater* (inner layer) extends into sulci and fissures, being adherent to the underlying neural tissue. The *arachnoid mater* (middle layer) lines the deep surface of the dura mater forming a network of fine threads bridging the pia and dura mater. The *dura mater* (outer layer) attaches to the inner surface of the skull, fusing with the endosteum: in places it separates from the adjacent bone forming large folds (*falx cerebri, tentorium cerebelli*) that project into the cranial cavity. Each fold consists of two layers and has a free edge.

The sickle-shaped falx cerebri projects from the roof of the cranial cavity in the median plane extending into the longitudinal fissure. The tentorium cerebelli lies in a transverse plane separating the occipital and temporal lobes from the underlying cerebellum. Posteriorly the falx cerebri fuses with the upper surface of the tentorium cerebelli.

The *subarachnoid space* (between arachnoid and pia mater) contains *cerebrospinal fluid*, which is reabsorbed into the bloodstream through extensions of the arachnoid mater (arachnoid granulations) that pierce the dura mater of the falx cerebri.

CRANIAL NERVES

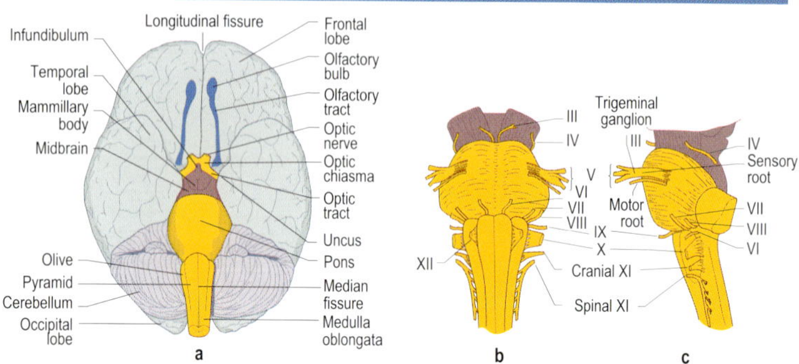

Position of cranial nerves related to (a) inferior surface of brain, (b) anterior and (c) lateral aspects of brainstem

Cranial nerves project from the brainstem and diencephalon connecting with structures and tissues in head, neck, thorax and abdomen. There are 12 pairs of cranial nerves known by number or name:

Olfactory nerve (I) – perception of smell
Optic nerve (II) – visual information from retina to optic chiasma
Oculomotor nerve (III) – innervates: superior, medial and inferior rectus; inferior oblique extraocular muscles: parasympathetic fibres to iris (pupil constriction)
Trochlear nerve (IV) – innervates superior oblique muscle
Trigeminal nerve (V) – three divisions (ophthalmic, maxillary, mandibular): mandibular division innervates muscles of mastication; all divisions sensory to skin of face
Abducens nerve (VI) – innervates lateral rectus muscle
Facial nerve (VII) – innervates muscles of facial expression: parasympathetic fibres stimulate crying and secretions from nasal, palatine, submandibular and sublingual glands
Vestibulocochlear nerve (VIII) – hearing and sense of balance
Glossopharyngeal nerve (IX) – sensory to pharynx and posterior third of tongue: sensory fibres innervate carotid body and sinus: parasympathetic fibres innervate parotid gland
Vagus (X) – innervates muscles of pharynx and larynx; parasympathetic fibres supply viscera in thorax and abdomen
Accessory nerve (XI) – cranial fibres join the vagus, spinal fibres innervate trapezius and sternomastoid
Hypoglossal nerve (XII) – innervates muscles of tongue

BLOOD SUPPLY: ARTERIES

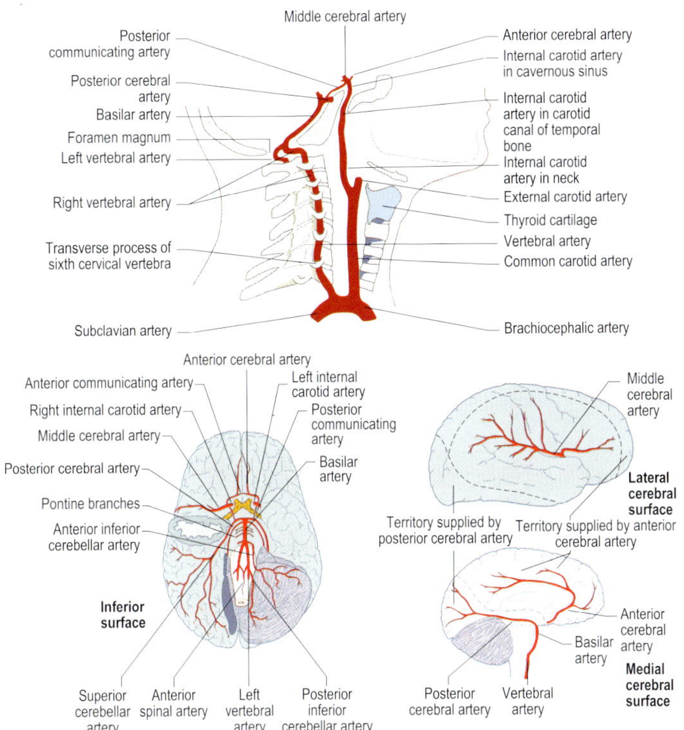

Arterial supply to the brain

The arterial supply is from paired *internal carotid* and *vertebral arteries*. The internal carotid passes through the carotid canal, along the foramen lacerum, through the cavernous sinus, piercing the dura and arachnoid mater to enter the subarachnoid space, where it divides into *anterior* and *middle cerebral arteries*. The anterior cerebral arteries supply the medial aspect of the cerebral hemispheres. The middle cerebral arteries supply the basal ganglia and external surface of the cerebral hemispheres, including the motor and sensory cortex.

The vertebral arteries pass through the foramina transversaria of C6 to C1, piercing the dura mater above C1, and then pass through the foramen magnum joining in the midline to form the *basilar artery* which supplies the medulla oblongata, pons and cerebellum. The terminal *posterior cerebral arteries* supply the midbrain and thalamus.

Each posterior cerebral artery anastomoses with the internal carotid via the *posterior communicating artery*. The ring of arteries formed is the *circle of Willis*.

BLOOD SUPPLY: VENOUS DRAINAGE

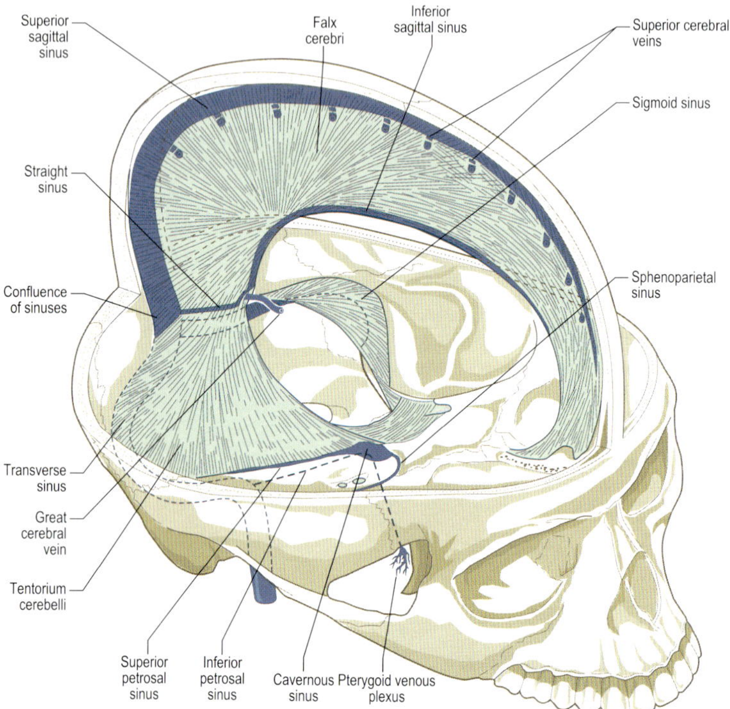

Dural venous sinuses of skull

The dural venous sinuses are located along the internal surface of the skull, most enclosed by bone externally and dura mater internally: others are located within specific folds of dura mater. The major venous sinuses are the single *superior* and *inferior sagittal* and *straight sinuses*, and the paired *cavernous, transverse* and *sigmoid sinuses*.

Internal aspects of the cerebral hemispheres, basal ganglia and thalamus are drained by the *great cerebral vein* to the straight sinus: the external surfaces drain towards the superior and inferior sagittal sinuses. The superior and straight sinuses drain into the transverse sinuses, which become the sigmoid sinuses and then the *internal jugular vein* as it passes through the jugular foramen. *Superior* and *inferior petrosal sinuses* drain the cavernous sinus into the sigmoid sinus and internal jugular vein respectively.

EAR

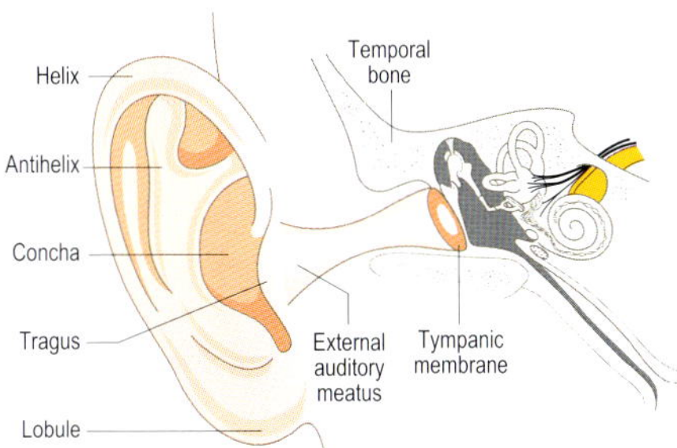

External, middle and internal ear

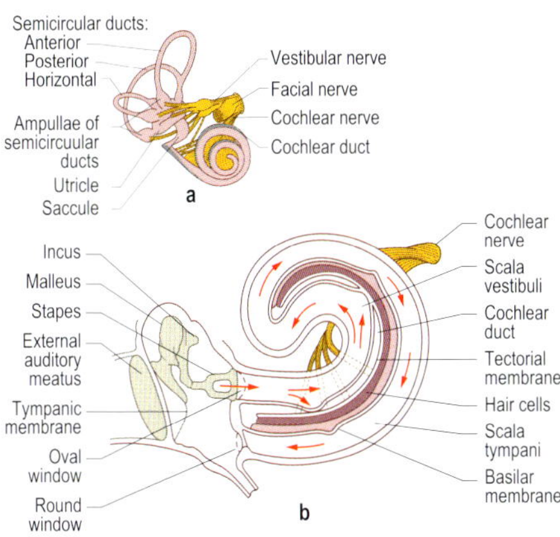

(a) Vestibular system; (b) conversion of sound waves into mechanical vibrations

The *external ear* collects and conveys sound towards the tympanic membrane, which separates it from the middle ear.

The *middle ear* contains the auditory ossicles (malleus, incus and stapes), which transmit vibrations of the tympanic membrane to the internal ear via the oval window. Vibration of the ossicles is damped by tensor tympani and stapedius muscles. The middle ear communicates with the nasopharynx by the auditory (Eustachian) tube, enabling equalization of pressure.

The *internal ear* is a complex series of fluid-filled spaces (membranous labyrinth) within the petrous temporal bone (bony labyrinth): displacement of fluid in these spaces stimulates sensory nerve endings. The bony labyrinth consists of three parts: the vestibule (utricle, saccule), the semicircular canals (anterior, posterior, horizontal) and the cochlea.

Semicircular canals convey information about rotatory and angular movements of the head, the utricle and saccule about linear and tilting movements.

The spiral cochlea is partly divided by the bony modiolus into the scali vestibuli above and scali tympani below. Movements of the stapes in the oval window cause pulsations of fluid in the scali tympani which are transmitted to the scali vestibuli. This causes movement of the basilar membrane and stimulation of the hair cells of the spiral organ resulting in auditory perception. Low-frequency sounds cause maximum activity in the basilar membrane; high-frequency sounds are limited to the basal part of the cochlea.

EYE

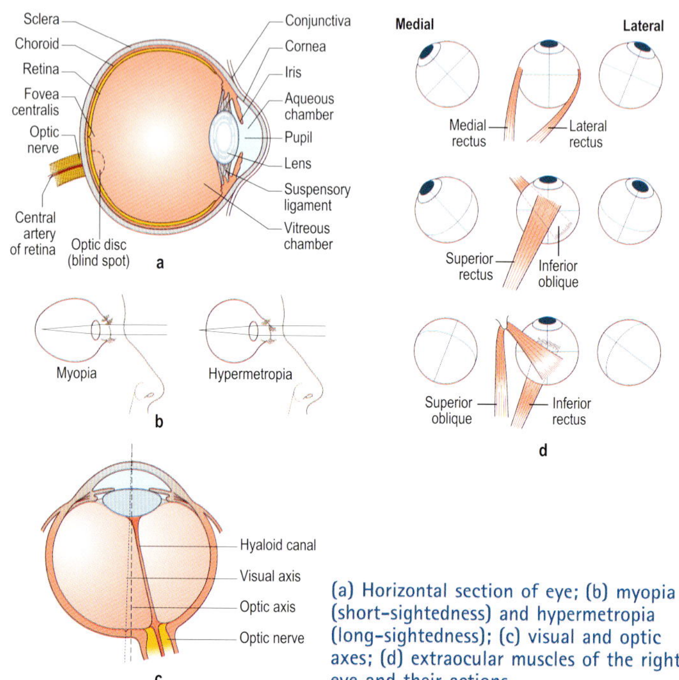

(a) Horizontal section of eye; (b) myopia (short-sightedness) and hypermetropia (long-sightedness); (c) visual and optic axes; (d) extraocular muscles of the right eye and their actions

The eyeball consists of an outer fibrous supporting layer (*sclera, cornea*), middle pigmented vascular layer (*choroid, ciliary body, iris*) and inner neural layer (*retina*), supported inferiorly by the suspensory ligament and protected by extraocular fat.

The optic axis is an imaginary line connecting the centre of the corneal curvature (anterior pole) to the centre of the scleral curvature (posterior pole). The visual axis joins the centre of the cornea and the fovea of the retina and represents the course taken by light from the centre point of vision. When looking at distant objects the visual axes of the two eyes are parallel.

Extraocular muscles (superior, medial, inferior and lateral rectus, and superior and inferior oblique) control the direction of gaze: medial and lateral rectus turn the eye to look horizontally medially and laterally; superior and inferior rectus pull medially in addition to turning it to look up and down; superior and inferior oblique pull the eye laterally as well as moving it down and up respectively. Extraocular muscles are supplied by the 3rd cranial nerve, except lateral rectus (6th cranial nerve) and superior oblique (3rd cranial nerve).

Index

C

F

G

H

I

J

M

N

O

Q

R

S

T

U

V

W

Z